T0262399

Advances in Polypropylene

Advances in Polypropylene

Edited by **Will Davis**

New York

Published by NY Research Press,
23 West, 55th Street, Suite 816,
New York, NY 10019, USA
www.nyresearchpress.com

Advances in Polypropylene
Edited by Will Davis

International Standard Book Number: 978-1-63238-038-8 (Hardback)

Printed in the United States of America.

Contents

Preface

The purpose of the book is to provide a glimpse into the dynamics and to present opinions and studies of some of the scientists engaged in the development of new ideas in the field from very different standpoints. This book will prove useful to students and researchers owing to its high content quality.

Polypropylene is widely used in the packaging and other industries. This book is a compilation of researches based on polypropylene, accomplished by experts around the globe over five decades. It demonstrates the various developmental stages that polypropylene has undergone. This book also consists of some current polypropylene theories. The book covers the various applications of polypropylene in varied industries. It intends to provide concise and well-compiled information related to polypropylene to its readers.

At the end, I would like to appreciate all the efforts made by the authors in completing their chapters professionally. I express my deepest gratitude to all of them for contributing to this book by sharing their valuable works. A special thanks to my family and friends for their constant support in this journey.

Editor

Polypropylene in the Industry

Polypropylene in the Industry of Food Packaging

Somaye Allahvaisi
Department of Entomology and Toxicology,
Faculty of Agriculture, Islamic Azad University of Tehran,
Branch of Sciences & Researches,
Iran

1. Introduction

Various pests expose agriculture and food products to attack from storage until consumption by consumers. Insects and fungi are the most serious pests that can contaminate food products in warehouses. Despite modern food and other agricultural products storage and distribution systems, most packaged food products, with the exception of canned and frozen goods, are subject to attack and penetration by insects (Mullen & Highland, 1988). When a packaging containing one of insect life stages enters into storages (infested packaging), it could cause the prevalence of infestation. In addition to reducing food quantity, insects annihilate quality, too. By nourishing into the foods, they prepare the conditions for the attack by pathogen microorganisms, such as fungi and as such, the consumption of these foodstuffs could be followed by dangerous present day diseases e.g. cancer types as contaminated foods to pathogens like fungi are one of the most important problems in the industry of storage foods and they are susceptible to mycotoxins (Jakic-Dimic et al., 2009). There are few categories of mycotoxins regarding their chemical structure, sensitivity of certain organs and origin of fungi that produce them. Aflatoxin is a secondary metabolite produced by *Aspergillus flavus* (Lopez-Diaz &Flannigan, 1997). Aflatoxin is potential to cause liver damage, cirrhosis, and liver cancer and aflatoxin B1 is the most dangerous toxin for animal and human health (Syarief et al., 2003). So, huge losses have been observed in agriculture produce and different ways are designed for controlling stored-product pests. Storing foodstuffs in bulk or sacks is a usual method for controlling pests without application of chemical methods. These sacks are made of different materials such as sheeted polymers used for packaging agricultural products to prevent the entrance of pests and contaminations (Allahvaisi, 2009). Wastage varies from 5-35% depending on nature of crops. Majority of wastage takes place in each of the steps viz. storage, transportation and at retail market due to improper packaging. Bulk Packaging made of polymers provides a solution for commodities weighing 10-50 kg during handling, storage and transportation, while smaller packaging for food products range from 50 ml to 5kg. Polymeric packaging fulfils the diverse role from protecting products, preventing spoilage, contamination, extending shelf life, ensuring safe storage thereby helping to make them readily available to consumers in our day to day life. This chapter will be a very helpful to all its readers, entrepreneurs, scientists, existing industries, technical institution, etc in the field of packaging (Anonymous, 2011).

2. Why plastics for packing?

Today, several polymer types are currently used for foodstuff packaging. Plastics have emerged as the most preferred choice of packaging materials for various products- from food, beverages, chemicals, electronic items and so on. They offer unique advantages over conventional materials (Anonymous, 2011):

- Safety: Plastics are safer materials for packaging of food products specially polyolefins which do not react with food. Pilferage and contamination is difficult.
- Shelf Life: Plastics packaging material offer better shelf life
- Cost: Plastics are the most cost effective medium of packaging when compared with any other material, the cost of transportation is reduced considerable on account of lower weight and less damage
- Convenience: Plastics can be converted in any form with various processing techniques, thus can pack any type of substances like liquids, powders, flakes, granules, solids.
- Waste: Packaging in plastics reduces the wastage of various food products, typical example is potatoes or onions packed in leno.
- Aesthetics: A right choice of plastics packaging increased the aesthetic value of products and helps in brand identity
- Handling and Storage: Products packed in plastics are easiest to handle and store as well as transport.
- Plastic products are easy to recycle.

Every day there are new products packed in plastics replacing conventional products and when a thought is given to pack a new product the first choice appears in the mind is Plastic packaging material.

3. Flexible plastic films

In general, flexible plastic films have relatively low cost and good barrier properties against moisture and gases; they are heat sealable to prevent leakage of contents; they add little weight to the product and they fit closely to the shape of the food, thereby wasting little space during storage and distribution; they have wet and dry strength, and they are easy to handle and convenient for the manufacturer, retailer and consumer. The main disadvantages are that (except cellulose) they are produced from non-renewable oil reserves and are not biodegradable. Concern over the environmental effects of non-biodegradable oil-based plastic packaging materials has increased research into the development of 'bioplastics' that are derived from renewable sources, and are biodegradable (Stewart, 1995). However, these materials are not yet available commercially in developing countries. There is a very wide choice of plastic films made from different types of plastic polymer. Each can have ranges of mechanical, optical, thermal and moisture/gas barrier properties. These are produced by variations in film thickness and the amount and type of additives that are used in their production. Some films (e.g. polyester, polyethylene, polypropylene) can be 'oriented' by stretching the material to align the molecules in either one direction (uniaxial orientation) or two (biaxial orientation) to increase their strength, clarity, flexibility and moisture/gas barrier properties. There are thus a very large number of plastic films and small-scale processors should obtain professional advice when selecting a material to ensure that it is suitable for the intended product and shelf life. Typically, the information required includes: type of plastic polymer(s) required; thickness/strength; moisture and gas permeability; heat seal temperature; printability on one or both sides; and suitability for use on the intended filling machinery (Ramsland, 1989; Robertson, 1993). Some may offer virtually no resistance against insects while others may be extremely resistant (Highland, 1981). Plastics based on Polypropylene (PP), polyethylene (PE), Polyvinylchloride (PVC) and Cellophane is mainly used for packaging applications (Table 1) (Odian, 2004).

Properties		Polyethylene	Polypropylene	Polyvinyl chloride	Cellophane
Max. heat tolerance (°C)		82-93	132-149	66-93	90-140
Min. heat tolerance (°C)		-57	-18	-46 to -29	-77
Sun light resistance		Moderate to good	moderate	good	good
Gas transmission (mm/100 cm² in 24 h and 25°C)	O_2	500	160	8-160	122-480
	N_2	180	20	1-70	33-90
	CO_2	2700	540	20-1900	2220
H_2O Absorption %		<0.01	<0.05	0	<0.03
H_2O Vapor transmission (g/100 cm² in 24h & 37.8°C & R.H. 90%)		1-1.5	0.25	4-10	0.2-1

Table 1. Some properties of used different polymers for packaging foodstuffs

A summary of the main different types of flexible plastic films is as follows (Anonymous, 2008):

3.1 Cellulose

Plain cellulose is a glossy transparent film that is odourless, tasteless and biodegradable (within approximately 100 days). It is tough and puncture resistant, although it tears easily. It has dead-folding properties that make it suitable for twist-wrapping (e.g. sugar confectionery). However, it is not heat sealable and the dimensions and permeability of the film vary with changes in humidity. It is used for foods that do not require a complete moisture or gas barrier, including fresh bread and some types of sugar confectionery. Cellulose acetate is a clear, glossy transparent, sparkling film that is permeable to water vapour, odours and gases and is mainly used as a window material for paperboard cartons (Chiellini, 2008).

3.2 Polyethylene (or polythene)

Low-density polyethylene (LDPE) is heat sealable, inert, odour free and shrinks when heated. It is a good moisture barrier but is relatively permeable to oxygen and is a poor odour barrier. It is less expensive than most films and is therefore widely used for bags, for coating papers or boards and as a component in laminates. LDPE is also used for shrink- or stretch-wrapping (see Technical Brief: Filling and Sealing Packaged Foods). Stretch-wrapping uses thinner LDPE (25 - 38 µm) than shrink-wrapping (45-75 µm), or alternatively, linear low-density polyethylene is used at thicknesses of 17 - 24 µm. The cling properties of both films are adjusted to increase adhesion between layers of the film and to reduce adhesion between adjacent packages (Fellows & Axtell, 2003). High-density polyethylene (HDPE) is stronger, thicker, less flexible and more brittle than LDPE and a better barrier to gases and moisture. Sacks made from HDPE have high tear and puncture resistance and have good seal strength. They are waterproof and chemically resistant and are increasingly used instead of paper or sisal sacks.

3.3 Polypropylene

Polypropylene is a clear glossy film with a high strength and puncture resistance. It has a moderate barrier to moisture, gases and odours, which is not affected by changes in humidity. It stretches, although less than polyethylene. It is used in similar applications to LDPE. Oriented polypropylene is a clear glossy film with good optical properties and a high tensile strength and puncture resistance (Bowditch, 1997). It has moderate permeability to gases and odours and a higher barrier to water vapour, which is not affected by changes in humidity. It is widely used to pack biscuits, snackfoods and dried foods (Hirsch, 1991).

3.4 Other films

Polyvinylidene chloride is very strong and is therefore used in thin films. It has a high barrier to gas and water vapour and is heat shrinkable and heat sealable. However, it has a brown tint which limits its use in some applications. Polyamides (or Nylons) are clear, strong films over a wide temperature range (from – 60 to 200°C) that have low permeability to gases and are greaseproof. However, the films are expensive to produce, require high temperatures to heat seal, and the permeability changes at different storage moistures. They are used with other polymers to make them heat sealable at lower temperatures and to improve the barrier properties, and are used to pack meats and cheeses (Paine & Paine, 1992).

3.5 Coated films

Films are coated with other polymers or aluminium to improve their barrier properties or to impart heat sealability. For example a nitrocellulose coating on both sides of cellulose film improves the barrier to oxygen, moisture and odours, and enables the film to be heat sealed when broad seals are used. Packs made from cellulose that has a coating of vinyl acetate are tough, stretchable and permeable to air, smoke and moisture. They are used for packaging meats before smoking and cooking. A thin coating of aluminium (termed 'metallisation') produces a very good barrier to oils, gases, moisture, odours and light (Lamberti and Escher, 2007). This metallised film is less expensive and more flexible than plastic/aluminium foil laminates (Lamberti & Escher, 2007).

3.6 Laminated films

Lamination (bonding together) of two or more films improves the appearance, barrier properties or mechanical strength of a package (Ramsland, 1989).

3.7 Coextruded films

Coextrusion is the simultaneous extrusion of two or more layers of different polymers to make a film. Coextruded films have three main advantages over other types of film: they have very high barrier properties, similar to laminates but produced at a lower cost; they are thinner than laminates and are therefore easier to use on filling equipment; and the layers do not separate. There are three main groups of polymers that are coextruded:

- Low-density and high-density polyethylene, and polypropylene.
- Polystyrene and acrylonitrile-butadiene-styrene.
- Polyvinyl chloride.

Typically,a three-layer coextrusion has an outside layer that has a high gloss and printability, a middle bulk layer which provides stiffness and strength, and an inner layer which is suitable for heat sealing. They are used, for example, for confectionery, snack-foods, cereals and dried foods. Thicker coextrusions (75 - 3000 μm) are formed into pots, tubs or trays.

4. Polymer films for packaging foodstuffs

Polymeric films have the most application in industry and are used in many packaging applications specially greenhouse and agricultural. In agricultural products that is the important subject in packaging, there are specific products include cereal, spices, edible oils, drinking water, chocolate and confectionery, fruits and vegetables, marine products and many more. So, there are various food items those are effectively and economically packed in various types of plastic packaging materials.

4.1 Physical properties of polymers

Physical properties of polymers include the degree of polymerization, molar mass distribution, crystallinity, as well as the thermal phase transitions:

- Tg, glass transition temperature
- Tm, melting point (for thermoplastics).

A plastic film suitable for use in fabricating a trash bag must exhibit strong physical properties in order to resist internal and external stresses on the bag. Such a bag could also be suitable for use as a container for shipping goods. In addition to resisting stresses, it is highly advantageous if the plastic film is easily heat sealable in order to simplify the manufacturing operations for producing the bags. The heat sealed seams must be strong and be capable of resisting stresses tending to break the seams (Liu et al., 2004).

4.1.1 Packaging polymers for preventing penetration of pest insects

Although finished products can be shipped from production facilities uninfested, stored product insects can enter packaged goods during transportation, storage in the

warehouse, or in retail stores. As from storage to consumption by consumers, the agriculture products are exposed to attack by pest insects. Insects are the most serious pests
that can contaminate the food by penetration of products in warehouses. The packaging of
products is the last line of defense for processors against insect infestation of their
finished products. There are two types of insects that attack packaged products:
"penetrators", which are insects that can bore holes through packaging materials; and
"invaders", which are insects that enter packages through existing holes, such as folds
and seams and air vents (Highland, 1984; Newton, 1988). The most insects use their sense
of olfaction to find food. The foodstuffs packages are made of different materials such as
sheeted polymers which are used for packaging the agricultural products in order to
prevention of entrance of pests. Consumer-size food packages vary considerably in their
resistance to insects. Sometimes the contamination was created by entrance of one
infested package. When neglected, such an infestation will serve as a source of infestation
for other commodities in the storage area. So, the packaging polymers should not only be
resistance to insects, but also should be permeable to gases used for disinfecting in stores.
Thus, the polymer thickness and manner of placing packages in storage should be
corrected to prevent serious damage in the products (Cline, 1978). Although, the
polymer`s kind is more important than thickness. In a study determined that the
difference between thicknesses of 16.5 and 29 µm is significant (Fig. 1). This figure shows
that the ability of species to penetrate materials may vary between life stages (Allahvaisi,
2010).

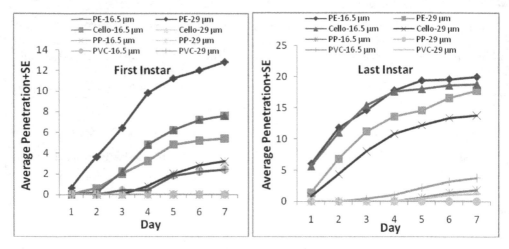

(PE=polyethylene, PP=polypropylene, PVC=polyvinylchloride and Cello=cellophane)

Fig. 1. Number of first and last instar larvae of *S. cerealella* that penetrated tested polymeric
pouches with two thick in lack of food conditions during 7-d period

As, remaining constant and subsequently decreasing the slope of the curves at insects'
penetration last days (after maximum penetration) prove that insects always attempt to
penetrate new food packages and their high activity is for availability to more food sources.
In bottom table you see the permeability percentage of four current polymers for packaging
foodstuffs in two thicknesses to some stored-pest insects starved.

	Pest insecs`s penetration in polymeric packagings (Average±SE)							
Polymer	Polyethylene		Cellophane		Polyvinyl chloride		Polypropylene	
Thickness (µm)	16.5	29	16.5	29	16.5	29	16.5	29
T. castaneum — A	12±0.45	4.2±0.36	13.8±0.5	6.4±0.76	3.6±0.4	0.0±0.0	0.0±0.0	0.0±0.0
1	a	b	a	c	c	d	d	d
F	0.0±0.0	0.0±0.0	0.0±0.0	0.0±0.0	0.0±0.0	0.0±0.0	0.0±0.0	0.0±0.0
L	11.45±0.0	3.8±0.36	10.6±0.4	3.4±0.4	0.0±0.0	0.0±0.0	0.0±0.0	0.0±0.0
2								
O. surinamensis — A	4±0.54	0.0±0.0	0.0±0.0	0.0±0.0	0.0±0.0	0.0±0.0	0.0±0.0	0.0±0.0
4								
C. maculates — A	9±0.31	2.4±0.22	6.2±0.36	5.4±0.22	0.0±0.0	0.0±0.0	0.0±0.0	0.0±0.0
3								
T. granarium — A	0.0±0.0	0.0±0.0	0.0±0.0	0.0±0.0	0.0±0.0	0.0±0.0	0.0±0.0	0.0±0.0
F	15.6±0.5	9.2±0.5	12.2±0.5	3.6±0.22	2.4±0.24	0.0±0.0	0.0±0.0	0.0±0.0
1	a	c	b	d	e	f	f	f
L	20±0.5	19.2±0.4	20±0.002	18.2±0.4	10±0.45	2.4±0.7	3.6±0.93	0.0±0.0
1	a	a	a	a	b	c	c	d
B. amydraula — F	8.8±0.36	3.6±0.22	5.8±0.36	5.8±0.36	1.8±0.2	0.0±0.0	0.0±0.0	0.0±0.0
1	b	d	a	d	c	e	d	f
L	20±0.002	18.02±0.32	19.4±0.22	18.6±0.4	7.6±0.22	5.4±0.22	10.4±0.22	3.4±0.22
1	a	a	a	a	b	d	c	e

([1]: Being Bilateral Effect and Duncan's Test Grouping , [2]: Disbilateral Effect of Polymer and Thickness , A: Adult, F: First Instar Larvae , L: Last Instar Larvae , [3]: Being Bilateral Effect , [4]: T-value)

Table 2. Average permeability percentage of different polymers to major stored-product insects in state of without food

Packaging polymers with repellents for preventing penetration of pest insects

In addition to improving the packaging material and design, insect repellents are used to prevent insects from entering packages by modifying the behavior of insects (Highland, 1984; Mullen, 1994; Watson and Barson, 1996; Mullen and Mowery, 2000). Pyrethrins synergized with piperonyl butoxide were approved for use as a treatment for insect-resistant packaging on the outer layer of packages or with adhesive in the USA (Highland, 1991). The repellency of pyrethrins was the primary mode of action against insect penetration and invasion (Laudani & Davis, 1955). Methyl salicylate, an insect repellent, has been registered to be used in food packaging to control stored-product insects in the USA (Radwan & Allin, 1997). DEET, neem, and protein-enriched pea flour are repellent to manystored-product insects when tested by exposure on filter paper or in preference chambers (Khan & Wohlgemuth, 1980; Xie et al., 1995; Fields et al., 2001). Included in the construction of the multiple-wall bags was a barrier layer that prevented the migration of repellents into the foodstuffs. So, a resistant polymer to insect`s penetration with a repellent of pests is the most suitable cover for packaging because it can prevent insect penetration and can be as a safe method for IPM programs which could in further reduce the application of the synthetic chemical pesticides and prevent the infestation of the stored-product pests. In some researches polypropylene polymer films are introduced as a suitable polymer with repellent for controlling the pest insects of stored-products. Research performed by Hou and colleagues (2004) showed that the repellents such as DEET reduce the number of insects entering the envelopes (Table 3).

| Insect | Number of insects | | χ^2 | P |
	Treated	Untreated		
Sitophilus oryzae	270.3	18973	735.82	<0.0001
Tribolium castaneum	1773	10178	197.88	<0.0001
Cryptolestes ferrugineus	470.7	117728	354.1	<0.0001
Oryzaephilus surinamensis	1173	10078	246.77	<0.0001
All insects	3475	507731	1481.56	<0.0001

Table 3. Number of insects (±SEM) in envelopes treated with DEET at 50 ml/envelope, 1 week after insects were released (n = 4)

4.2 Chemical properties of polymers

The attractive forces between polymer chains play a large part in determining a polymer's properties. Because polymer chains are so long, these interchain forces are amplified far beyond the attractions between conventional molecules. Also, longer chains are more amorphous (randomly oriented). Polymers can be visualised as tangled spaghetti chains - pulling any one spaghetti strand out is a lot harder the more tangled the chains are. These stronger forces typically result in high tensile strength and melting points. The intermolecular forces in polymers are determined by dipoles in the monomer units. Polymers containing amide groups can form hydrogen bonds between adjacent chains; the positive hydrogen atoms in N-H groups of one chain are strongly attracted to the oxygen atoms in C=O groups on another. These strong hydrogen bonds result in, for example, the high tensile strength and melting point of Kevlar (Anonymous, 2011).

4.2.1 Polymers and permeability to fumigants for controlling pests through packages

In spite of the advances recorded in many aspects of stored product pest control, fumigation being a no residual chemical treatment has remained the mainstay for control of stored product pests. Therefore, it is accepted that fumigation is the most universal and the less hazardous method for maintaining of agricultural products under storage conditions (Keita et al., 2001).

Frequently products are packed in jute bags or plastic bags. Since penetration of the fumigant into the bags is a critical factor it is evident that fumigations under tarps or plastic sheeting should take into account the properties of the packaging materials. For controlling the insect pests by fumigants, the gas must penetrate from the air-space beneath the tarps into the bags containing the stored products. The passage of gas through these polymers to lower layers for eradicating the contamination into packaged foodstuffs is one of the other goals of storage in long-times. Polymers with various thicknesses have different permeability to fumigants (Stout, 1983; Appert, 1987; ACIAR, 1989; Iqbal et al., 1993; Valentini, 1997; Hall, 1970; Marouf & Momen, 2004) (Table 4). So, determining the best thickness of polymer is important in packaging for controlling pests. By incomplete fumigation; specially quarantine pests into packagings can easily enter countries within packaged products. In certain cases such as dried fruit, which are packed in plastic bags, entrance of fumigant into the bags is critical in controlling stored-product insect pests that originate in the field. Some studies are evident that polypropylene liners of less than 100µm thickness are suitable as inner liners of jute bags to allow the fumigant to enter the bags (because of their high permeability) (Fleural – Lessard & Serrano, 1990; Sedlacek, 2001).

	CO2 gas insecs`s penetration in polymeric packagings (Average±SE)							
Polymer	Polyethylene		Cellophane		Polyvinyl chloride		Polypropylene	
Thickness (µm)	16.5	29	16.5	29	16.5	29	16.5	29
	1.3±0.013	0.44±0.004	1.28±0.01	0.443±0.005	0.4±0.012	0.23±0.004	0.73±0.004	0.32±0.007
	a	b	a	c	d	f	c	e

Table 4. The tested polymers to mean permeability the polymers to CO2 gas

4.2.2 Antimicrobial polymers

The subject is covalently bonding anti-microbial agents to the surface of a selected polymer and its method of use as an anti-microbial agent to reduce surface bacterial, fungus, and/or virus count of the material it contacts (Jo et al., 2009). This can be applied to a variety of applications such as film and container packaging of foodstuffs, cosmetics, medical equipment and devices, environmental, hygienic and sanitary applications, as well as other consumer and commercial use (Kenaway et al., 2007). So, the applied polymers for packaging should be have the ability of coating to materials like nano metals such as silver nanoparticles to gain antimicrobial properties because in usual conditions, there is much growth of microbe agents on packaging polymers (Fig. 2).

Edible coatings have long been known to protect perishable food products from deterioration by retarding dehydration, suppressing respiration, improving textural quality, helping retain volatile flavor compounds and reducing microbial growth (Debeaufort et al., 1998). Specially formulated edible coatings may provide additional protection against contamination of microorganism while serving the similar effect as modified atmosphere storage in modifying internal gas composition (Park, 1999). Among noble-metal nanomaterials, silver nanoparticles have received considerable attentions due to their attractive physicochemical properties. It is well known that silver in various chemical forms has strong toxicity to a wide range of microorganisms (Liau et al., 1997). The larger surface area of silver nanoparticles can improve their antibacterial effectiveness against 150 types of microbes. Although the coating has been extensively studied to increase the shelf life of many agricultural products, little information is available regarding the application of silver nanoparticles-polymers coating for these products.

On the other hand, antimicrobial Polymers, known as polymeric biocides, are a class of polymers with antimicrobial activity, or the ability to inhibit the growth of microorganisms such as bacteria, fungi or protozoans (Fig. 3). In this figure, normal bacterial membranes (panel a) are stabilized by Ca+2 ions binding anionic charged phospholipids. NIMBUS™ quat-polymer rapidly displaces Ca+2 (panel b) leading to loss of fluidity (panel c) and eventual phase separation of different lipids. Domains in the membrane then undergo a transition to smaller micelles. These polymers have been engineered to mimic antimicrobial peptides which are used by the immune systems of living things to kill bacteria. Typically, antimicrobial polymers are produced by attaching or inserting an active antimicrobial agent onto a polymer backbone via an alkyl or acetyl linker. Antimicrobial polymers may enhance the efficiency and selectivity of currently used antimicrobial agents, while decreasing associated environmental hazards because antimicrobial polymers are generally nonvolatile and chemically stable. This makes this material a prime candidate for use in areas of medicine as a means to fight infection, in the food industry to prevent bacterial contamination, and in water sanitation to inhibit the growth of microorganisms in drinking water (Pyatenko et al., 2004).

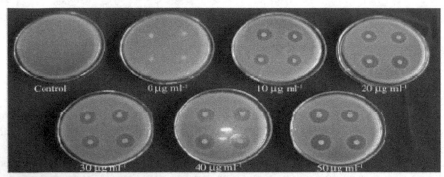

a) Inhibition zone caused by AgNO₃

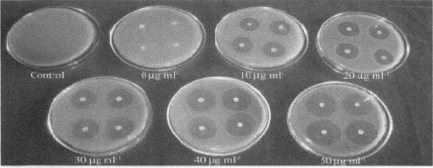

b) Inhibition zone caused by silver nanoparticles

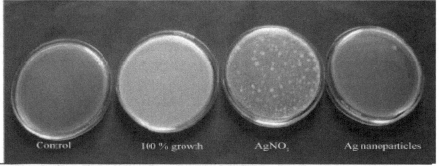

c) Population of *E. coli* at 50 mg ml⁻¹ of AgNO₃ and silver nanoparticles

Fig. 2. Comparison of E.coli growth inhibition by AgNO3 and silver nanoparticles (Parameswari et al., 2010)

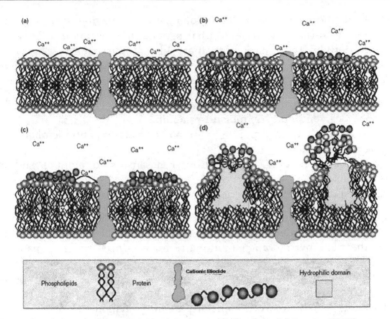

Fig. 3. Action of polymeric cationic biocidal agent (Gilbert and Moore, 2005)

4.2.3 Edible films and coatings

An edible film or coating is simply defined as thin continuous layer of edible material formed on, placed on or between foods or food components (Torres, 1994). Edible packaging refers to the use of edible films, coatings, pouches, bags and other containers as a means of ensuring the safe delivery of food product to the consumer in a sound condition (McHugh & Krochta, 1994). These films can also act as carrier of antioxidant, flavour and bacteriostats and can improve mechanical integrity of food products (Pathania, 2000). Since, package is an integral part of the whole food product, therefore, the composition of the edible packaging must meet with the following specific functional requirements:

- Neutral organoleptic properties (clear, transparent, odourless, tasteless etc.)
- Water vapour tightness to prevent desiccation.
- Good barrier against microbial invasion to reduce spoilage and decay.
- Predetermined permeability to water vapour, O_2 and CO_2 to have complete control over the water and gas exchanges between the product and surrounding atmosphere.
- Good mechanical characteristics (like tensile and yield strength, Spencer impact elongation, etc.) to impart abuse resistence.
- Enchance the surface appearance (e.g. brilliance) and tactile characteristics (e.g. reduced stickiness) of foods (Kaushik, 1999).

4.2.4 Nanotechnology applications in foodstuffs packaging polymers

The development of nanodevices and nanomaterials could open up novel applications in agriculture (Scrinis & Lyons, 2007). Nanophasic and nanostructured materials are attracting a great deal of attention because of their potential for achieving specific processes and selectivity, especially in biological and pharmaceutical applications (Pal et al., 2007). Nanotechnology derived food packaging materials are the largest category of current nanotechnology applications for the food sector. The main applications for food contact materials (FCMs) including:

1. FCMs incorporating nanomaterials to improve packaging properties (flexibility, gas barrier properties, temperature/moisture stability).
2. "Active" FCMs that incorporate nanoparticles with antimicrobial or oxygen scavenging properties.
3. "Intelligent"food packaging incorporating nanosensors to monitor and report the condition of the food.
4. Biodegradable polymer–nanomaterial composites (Chaudhry et al., 2008).

Polymer composites are mixtures of polymers with inorganic or organic fillers with certain geometries (fibers, flakes, spheres, particulates). The use of fillers which have at least one dimension in the nanometric range (nanoparticles) produces polymer nanocomposites. Three types of fillers can be distinguished, depending on how many dimensions are in the nanometric range. Isodimensional nanoparticles, such as spherical silica nanoparticles or semiconductor nanoclusters, have three nanometric dimensions. Nanotubes or whiskers are elongated structures in which two dimensions are in the nanometer scale and the third is larger. When only one dimension is in the nanometer range, the composites are known as polymer-layered crystal nanocomposites, almost exclusively obtained by the intercalation of the polymer (or a monomer subsequently polymerized) inside the galleries of layered host crystals (Azeredo, 2009). There are three common methods used to process nanocomposites: solution method, *in situ* or interlamellar polymerization technique, and melt processing. The solution method can be used to formboth intercalated and exfoliated nanocomposite materials. In the solution method, the nanocomposite clay is first swollen in a solvent. Next, it is added to a polymer solution, and polymer molecules are allowed to extend between the layers of filler. The solvent is then allowed to evaporate. The *in situ* or interlamellar method swells the fillers by absorption of a liquid monomer. After the monomer has penetrated in between the layers of silicates, polymerization is initiated by heat, radiation, or incorporation of an initiator. The melt method is the most commonly used method due to the lack of solvents. In melt processing, the nanocomposite filler is incorporated into a molten polymer and then formed into the final material (Brody et al., 2008). The results of tests performed by An and colleagues (2008) shown in Fig. 4 revealed the evidence for the formation of silver nanoparticles in the coating solutions prepared under the experimental condition. The solutions with PVP formed a thin coating on the surface of asparagus when water evaporated, leaving the nanoparticles evenly distributed in the coating matrix (Jianshen et al., 2008).

Nanocomposite packages are predicted to make up a significant portion of the food packaging market in the near future. Silver is well known for its strong toxicity to a wide range of microorganisms (Liau et al., 1997), besides some processing advantages such as high temperature stability and low volatility (Kumar & Münstedt, 2005). Silver nanoparticles have been shown to be effective antimicrobials (Aymonier et al., 2002; Sondi & Salopek-Sondi, 2004; Son et al., 2006; Yu et al., 2007; Tankhiwale & Bajpai, 2009), even more effective than larger silver particles, thanks to their larger surface area available for interaction with microbial cells (An et al., 2008; Kvítek et al., 2008). In fact, the most common nanocomposites used as antimicrobial films for food packaging are based on silver nanoparticles, whose antimicrobial activity has been ascribed to different mechanisms, namely: (a) adhesion to the cell surface, degradation of lipopolysaccharides and formation of "pits" in the membranes, largely increasing permeability (Sondi & Salopek-Sondi, 2004); (b) penetration inside bacterial cell, damaging DNA (Li et al., 2008); and (c) releasing antimicrobial Ag+ ions by dissolution of silver nanoparticles (Morones et al., 2005). The latter mechanism is consistent with findings by Kumar & Münstedt (2005), who have concluded that the antimicrobial activity of silverbased systems depends on releasing of Ag+, which binds to electron donor groups in biological molecules containing sulphur, oxygen or nitrogen. Besides the antimicrobial activity, silver nanoparticles have been reported to absorb and decompose ethylene, which may contribute to their effects on extending shelf life of fruits and vegetables (Li et al., 2009).

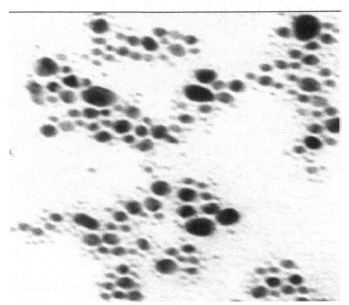

Fig. 4. Transmission electron microscopy (TEM) of silver nanoparticles (×100,000) (An et al, 2008)

5. Suitable polymer for stored-product packaging

Plastics based on Polypropylene, Polyethylene, Polyvinyl Chloride and Cellophane, hugely used for packagings, has some of these properties but this is different at them. For example, these polymers rank generally from the easiest to the most difficult to penetration against insect pests; Cellophane, polyethylene, Polyvinylchloride and Polypropylene. The least penetration is carried out in PP and PVC polymers. Foodstuffs packaged by polymer films of PP and PVC could provide the conditions and so, by suitable packaging the stored pest insects do not access to food and without food they become extinct. But in the comparison between polypropylene and polyvinylchloride, PVC isn't a safe polymer for packaging foodstuffs in order to release HCl gas and the only importance of PVC in storage industry is often to be used as a gas-tight cover on agricultural products to keep a suitable concentration of gas and it is important for controlling quarantine pests. Furthermore, other two polymers, Polyethylene and Cellophane have a high permeability to gases but a very low resistance to pests as the product packaged into them becomes more contamination than ones into PVC and PP. The polymer films of Polyethylene and Cellophane; specially Cellophane, is greatly used for packaging the products be consumed daily. Moreover, Cellophane is 100% biodegradable. Some studies show that polypropylene had a good degradability in environment in comparative to polyethylene and polyvinylchloride. Also, new studies show that polypropylene has a suitable property for coating with nano metals and repellents for decreasing the losses of stored-products in effect of pest infestation. Hence, according to the investigations of researchers (in above) polypropylene usually is used as a suitable cover for packaging foodstuffs in stores and has perfect physical and chemical properties for the packaging works which should be performed in stores of maintaining foodstuffs.

6. Polypropylene as packaging polymer

PP known as polypropene, is one of those most versatile polymers available with applications, both as a plastic and as a fibre, in virtually all of the plastics end-use markets. Professor Giulio Natta produced the first polypropylene resin in Spain in 1954. Natta utilised catalysts developed for the polyethylene industry and applied the technology to propylene gas. Commercial production began in 1957 and polypropylene usage has displayed strong growth from this date. PP is a linear hydrocarbon polymer, expressed as

Polypropylene	
IUPAC name poly(propene)	
Other names Polypropylene; Polypropene; Polipropene 25 [USAN];Propene polymers; Propylene polymers; 1-Propene	
Identifiers	
CAS number	9003-07-0
Properties	
Molecular formula	$(C3H6)_n$
Density	0.855 g/cm3, amorphous 0.946 g/cm3, crystalline
Melting point	130–171°C
(what is this?) (verify) Except where noted otherwise, data are given for materials in their standard state (at 25°C, 100 kPa)	

C_nH_{2n}. PP, like polyethylene (see HDPE, L/LLDPE) and polybutene (PB), is a polyolefin or saturated polymer. (Semi-rigid, translucent, good chemical resistance, tough, good fatigue resistance, integral hinge property, good heat resistance). PP does not present stress-cracking problems and offers excellent electrical and chemical resistance at higher temperatures. While the properties of PP are similar to those of Polyethylene, there are specific differences. These include a lower density, higher softening point (PP doesn't melt below 160°C, Polyethylene, a more common plastic, will anneal at around 100°C) and higher rigidity and hardness (Cacciari, 1993). Additives are applied to all commercially produced polypropylene resins to protect the polymer during processing and to enhance end-use performance. PP is a thermoplastic which is commonly used for plastic moldings, stationary folders, packaging materials, plastic tubs, non-absorbable sutures, diapers etc. PP can be degraded when it is exposed to ultraviolet radiation from sunlight. Furthermore, at high temperatures, PP is oxidized. The possibility of degrading PP with microorganisms has been investigated.

Three main types of PP polymer types are used in household packaging:

1. Homopolymer PP: this is a translucent polymer, with high Heat Distortion Temperature (HDT), with a lower impact strength (particularly at low temperatures) and is used for applications such as closures and soup pots;
2. Block copolymer PP: this polymer has a lower transparency and generally a lower HDT, with a higher impact strength (particularly at low temperatures) and is used for applications such as ice cream containers and for chilled foods;
3. Random copolymer PP: this polymer has a high transparency and the lowest HDT. It is a product with the greatest flexibility and possesses reasonable impact strength. Typical applications requiring high transparency are bottles and salad bowls; Homopolymer and copolymer (random and block) PP polymer types may be used with either of the two main types of moulding process (extrusion/thermoforming or extrusion blow moulding) and therefore can be made with different melt flow characteristics as follows:
4. Thermoforming and blow moulding: used for meat trays and bottles, with a low MFR (Melt Flow Rate) (1 to 4);
5. Injection moulding: used for thin walled packaging, such as soup pots, with a high MFR (33 and higher).

Thus, Polypropylene is a thermoplastic polymer used in a wide variety of applications including packaging, textiles (e.g. ropes, thermal underwear and carpets), stationery, plastic parts and reusable containers of various types, laboratory equipment, loudspeakers, automotive components, and polymer banknotes. An addition polymer made from the monomer propylene, it is rugged and unusually resistant to many chemical solvents, bases and acids. The versatility of the polymer (the ability to adapt to a wide range of fabrication methods and applications) has sustained growth rates enabling PP to challenge the market share of a host of alternative materials in plethora of applications. In 2007, the global market for polypropylene had a volume of 45.1 million tons, which led to a turnover of about 65 billion US-dollars (47.4 billion Euro) (Kenaway et al., 2007). PP films are used in household paper packing, stationery packaging, portfolios and food packaging. In stationer industry, they are used for photo albums and page protectors. PP films are used as a lamination layer, both as sealant and as heat resistant layer and in the pressure sensitive industry for adhesive coating and diaper closures. Various packaging of products could be made of this film.

7. Conclusion

According to the results of performed works in the field of packaging, it is proved that a polymeric cover usually made of polypropylene with thickness <100 µm is the most suitable one for foodstuffs packaging. In the less thickness, some polymers have less resistant to the infestation of pest insects although polypropylene shows resistance well ever in lower thicknesses. Such cover would undoubtedly reduce the danger of crossinfestation and on the one hand, propylene is permeable to stored gases such as phosphine for ruining the contaminations into stored products and has the ability of coating to nano metals in the thickness and thus could obtain the antimicrobial properties. Also, propylene has a good degradability among polyolefines. A consultation exercise with the PP packaging supply chain explored the levels of interest of using a food grade PP in packaging applications. Little recycled PP is currently used in packaging because little is currently available. There is certainly interest from all levels of the supply chain – retailers, brand owners, food manufacturers and packaging manufacturers – in using recycled PP, if a recycling system existed that could meet regulatory standards and company food performance standards (Anonymous, 2010). Such a change would undoubtedly reduce the too using of chemical pesticides and increase the storages food maintaining and therefore reduce economic losses associated with infestation and minimize injury to company image as a manufacture of high quality foodstuffs.

8. Acknowledgment

I would like to thank Ms. H. Salimizand, Dr. A.A. Pourmirza and Dr. M.H. Safaralizade for considerable assistance. I also acknowledge younger researchers club of Sanandaj for some facilities.

9. References

[1] Aciar, A. (1989). Suggested recommendations for the fumigation of grain in the ASEAN region, Part 1. *Principles and General Practice. 131 pp, ISSN 1832-1879.*

[2] Appert, J. (1987). The storage of food grains and seeds. *Macmillan. London, 146 pp, ISBN 13: 9780333448274.*

[3] Allahvaisi, S.; Pourmirza, A.A. & Safaralizade, M.H. (2009). Packaging of Agricultural Products for Preventing Tobacco Beetles Contaminations. *Notul. B. Horti Agrobo. Cluj-N. 37(2): 218-222, ISSN 0255-965X.*

[4] Allahvaisi, S.; Pourmirza, A.A. & Safaralizade, M.H. (2010). The study on polymers permeability for foodstuffs packaging by some serious species of stored pest insects and phosphine gas. *J. Agri. Tech. 6 (4): 747-759, ISSN 1680-7073.*

[5] An, J.; Zhang, M.; Wang, S. & Tang, J. (2008). Physical, chemical and microbiological changes in stored green asparagus spears as affected by coating of silver nanoparticles-PVP. *LWT – Food Science and Technology. 41(6): 1100– 1107, ISSN 0023-6438.*

[6] Anonymous. (2008). Packaging materials for food. *Technical Brief, Practical Action.*

[7] Anonymous. (2010). Scoping study into food grade polypropylene recycling. *Available in:* www.wrap.org.uk/foodgradepp,

[8] Anonymous. (2011). Plastics in food packaging in India.

[9] Anonymous. (2011). Physical and chemical properties of polymers. *Available in: Plastics.indiabizclub.com.*

[10] Aymonier, C.; Schlotterbeck, U.; Antonietti, L.; Zacharias, P.; Thomann, R. & Tiller, J.C. (2002). Hybrids of silver nanoparticles with amphiphilic hyperbranched macromolecules exhibiting antimicrobial properties. *Chemi. Commun. 24: 3018–3019, ISSN 1364-548X.*

[11] Azeredo, H.M.C. (2009). Nanocomposites for food packaging applications. *Food Res. Intern. 42(9): 1240-1253, ISSN 0963-9969, ISSN 0963-9969.*

[12] Bowditch, T.G. (1997). Penetration of Polyvinyl Choloride and Polypropylene Packaging Films by *Ephestia cautella* (*Lepidoptera: Pyralidae*) and *Plodia interpunctella* (*Lepidoptera: Pyralidae*) Larvae, and *Tribolium confusom* (*Coleoptera: Tenebrionidae*). *J. Eco. Entomol. 90(4):1028-101, ISSN 0022-0493.*

[13] Brody, A.; Bugusu, B.; Han, J.H.; Sand, C.K. & Machugh, T.H. (2008). Inovative food packaging solutions. *J. Food Sci. 73(8): 107 - 116, ISSN 1750-3841.*

[14] Cacciari, I.; Quatrini, P.; Zirletta, G.; Mincione, E.; Vinciguerra, V.; Lupatelli & P.; Sermanni, G.G. (1993). Isotactic polypropylene biodegradation by a microbial community: Physico-chemical characterization of metabolites produced. *Appl. Environ. Microbiol. 59: 3695-3700, ISSN 0099-2240.*

[15] Chaudhry, Q. (2008). "Applications and Implications of Nanotechnologies for the Food Sector." *Food Addi. and Contami. 25(3): 241–258, ISSN 1464–5122.*

[16] Chiellini, E. (2008). Environmentally-compatible Food Packaging. *Woodhead Publishing, Cambridge, ISSN 1811-5209.*

[17] Cline, L.D. (1978). Penetration of seven common packaging materials by larvae and adults of eleven species of stored-product insects. *J. Eco. Entomol. 71:726-729, ISSN 0022-0493.*

[18] Debeaufort, F.; Quezada-Gallo, J.A. & Voilley, A. (1998). Edible films and coatings: Tomorrow's packaging: A review. *Critical revi. In Food Sci. 38:299–313, ISSN 1750-3841.*

[19] Fellows, P.J. & Axtell, B.L. (2003). Appropriate Food Packaging: Materials and methods for small businesses. *2nd Edn. Practical Action Publishing, ISSN 0254-6019.*

[20] Fields, P.G; Y.S., Xie & Hou, X. (2001). Repellent effect of pea (Pisum sativum) fractions against stored-product insects. *J. Stored Pro. Res. 37: 359–370, ISSN 0022-474X.*

[21] Fleural – Lessard, F. & Serrano, B. (1990). Resistance to insect perforation of plastic films for stored – product packing, "methodological study on tests with rice weevil and layer grain borer. *Sci. Aliments. 10(3): 521-532, ISSN 0240-8813.*

[22] Gilbert, P. and L. E., Moore. 2005. Cationic Antiseptics: diversity of action under a common epithet, *J. Appl. Microbiol. 99(4):703-15, ISSN 1365-2672.*

[23] Hall, D.W. (1970). Handling and storage of food grains in tropical and subtropical areas. *FAO, Rome, 350 pp, ISSN 1045-7127.*

[24] Highland, H.A. (1975). Insect-resistant textile bags: new construction and treatment techniques. *USDA Technical Bulletin, pp.1511, ISSN 0082-9811.*

[25] Highland, H.A. (1981). Resistant barriers for stored- product insects. In: CRC Handbook of transportation and marketing in agriculture. *Vol. 1: Food commodities, 41-45, ISSN 0817-8038.*

[26] Highland, H.A. (1984). Insect infestation of packages, 311-320 pp. In: F. J. Baur (Eds.). Insect Management for Food Processing. *American Association of Cereal Chemists, St. Paul, MN, ISSN 0714-6221.*

[27] Highland, H.A.; Kamel, A.H.; Sayed, M.M.EL.; Fam, E.Z.; Simonaitis, R. & Cline, L.D. (1984). Evaluation of permethrin as an insect-resistant treatment on paper bags and of tricalcium phosphate as a suppressant of storedproduct insects. *J. Eco. Entomol. 77: 240–245, ISSN 1938-291X.*

[28] Highland, H.A. (1991). Protecting packages against insects. In: Gorham, J.R. (Ed.), Ecologyand Management of Food-Industry Pests. *Association of Official Analytical Chemists, Arlington, VA, pp. 345–350, ISSN 1516-8913.*

[29] Hirsch, A. (1991). Flexible Food Packaging. *Van Nostrand Reinhold, New York, ISSN 1092-3659.*

[30] Hou, X.; Fields, P. & Taylor, W. (2004). The effect of repellents on penetration into packaging by stored-product insects. *J. Stored Prod. Res. 40: 47–54, ISSN 0022-474X.*

[31] Iqbal, J.; Irshad, & Khalil, S.K. (1993) Sack fumigation of wheat under polythene sheets. *Sarhad J. Agri. IX (5): 399-402, ISSN 1016-4383.*

[32] Jakic-Dimic, D.; Nesic, K. & Petrovic, M. (2009): Contamination of cereals with aflatoxins, metabolites of fungi *Aspergillus flavus*. Biotech. *In Animal Husba. 25: 1203–1208, ISSN 0378-4320.*

[33] Jianshen A.; Min, Zh.; Shaojin, W. & Juming, T. (2008). Physical, chemical and microbiological changes in stored green asparagus spears as affected by coating of silver nanoparticles-PVP. *J. scin. dir. 41:1100–1107, ISSN.*

[34] Jo, Y.; Kim, B. & Jung, G. (2009). Antifungal activity of silver ions and nanoparticles on phytopathogenic fungi. *Plant Disease. 93:1037-1043, ISSN 0191-2917.*

[35] Khan, M.A. & Wohlgemuth, R. (1980). Diethyltoluamide as a repellent against stored-products pests. *Anzeiger fur Schadlingskunde Pflanzenschutz Umweltschutz 53, 126–127, ISSN 1439-0280.*

[36] Keita, S.M.; Vincent, C.; Schmit, J.; Arnason, J.T. & Belanger, A. (2001). Efficacy of essential oil of *Ocimum basilicum* L. and *O. gratissimum* L. applied as an insecticidal fumigant and powder to control *Callosobruchus maculatus* (Fab.) (Coleoptera: Bruchidae). *J. Stored Prod. Res. 37: 339-349, ISSN 0022-474X.*

[37] Kenaway, El-Refaie; Worley, S.D. & Roy, B. (2007). "The Chemistry and Applications of Antimicrobial Polymers: A State of the Art Review". BioMacromolecules (American Chemi. Soc.). 8 (5): 1359–1384, ISSN 0002-7863.

[38] Kumar, R. & Munstedt, H. (2005). *Silver ion release from antimicrobial polyamide/silver composites. Biomaterials. 26: 2081– 2088, ISSN 0142-9612.*

[39] Kvitek, L.; Panac_ek, A.; Soukupova, J.; Kolar˘, M.; Vec_er˘ova, R. & Prucek, R. (2008). Effect of surfactants and polymers on stability and antibacterial activity of silver nanoparticles (NPs). *J. Physi. Chemi. C. 112(15): 5825–5834, ISSN 1932-7455.*

[40] Lamberti, M. & Escher, F. (2007). Aluminium foil as a food packaging material in comparison with other materials. *Food Rev. Intern. 23 (4): 407-433, ISSN 8755-9129.*

[41] Laudani, H. & Davis, D.F. (1955). The status of federal research on the development of insect-resistant packaging. *TAPPI. 38: 322–326, ISSN 0734-1415.*

[42] Li, H.; Li, F.; Wang, L.; Sheng, J.; Xin, Z.; Zhao, L.; Xiao, H.; Zheng, Y. & Hu, Q. (2009). Effect of nano-packing on preservation quality of Chinese jujube (*Ziziphus jujuba* Mill. var. inermis (Bunge) Rehd). *Food Chemistry, Vol. 114, No. 2, 547-552, ISSN 0308-8146.*

[43] Li, Q.; Mahendra, S.; Lyon, D. Y.; Brunet, L., Liga, M.V.; Li, D. & Alvarez, P.J.J. (2008). Antimicrobial nanomaterials for water disinfection and microbial control: potential applications and implications. *Water Research, 42(18): 4591– 4602, ISSN 0043-1354.*

[44] Liau, S.Y.; Read, D.C.; Pugh, W.J.; Furr, J.R. & Russell, A.D. (1997). Interaction of silver nitrate with readily identifiable groups: relationship to the antibacterial action of silver ions. *Letters in Applied Microbiology, 25: 279– 283, ISSN 1472-765X.*

[45] Liu, Ch.; Tellez-Garay, A.M. & Castell-Perez, M.E. (2004). Physical and mechanical properties of peanut protein films. J. Food Sci.and Techn. 37(7): *731-738, ISSN 1750-3841.*

[46] Marouf, A. & Momen, R.F. (2004). An evaluation of the permeability to phosphine through different polymers used for the bag storage of grain. *Int. Conf. Controlled Atmosphere and Fumigation in Stored Prod. Gold-Coast Australia, 8-13 August, FTIC Ltd. Publishing, Israel, ISSN 1021-9730.*

[47] McHugh, T.Habig; Avena-Bustillos, R.D. & Krochta, J.M. (1993). Hydrophilic edible films-modified procedure for water vapor permeability and explanation of thickness effects. *J. Food Sci. 58: 899–903, ISSN 1750-3841.*

[48] Morones, J.R.; Elechiguerra, J.L.; Camacho, A.; Holt, K.; Kouri, J.B. & Ramirez, J.T. (2005). The bactericidal effect of silver nanoparticles. *Nanotechnology. 16(10): 2346–2353, ISSN 1361-6528.*

[49] Mullen, M.A. & Highland, H.A. (1988). Package Defects and Their Effect on Insect Infestation of Infestation of Instant Non-fat Dry Milk. *J. Packag. Tech. 2:226-269, ISSN 1099-1522.*

[50] Mullen, M.A. & Mowery, S.V. (2000). Insect-resistant packaging. *Intern. Food Hyg. J. 11: 13–14, ISSN 0015-6426.*

[51] Newton, J. (1988). Insects and packaging—a review. Intern. Biodet. *J. 24: 175–187, ISSN 0964-8305.*

[52] Odian, G. (2004). Principles of polymerization. *USA. 4 edition, 832 pp.*

[53] Paine, F.A. & Paine, H.Y. (1992). A Handbook of Food Packaging, 2nd Edition, *Blackie Academic and Professional, London, ISSN 1532-1738.*

[54] Pal, S., Tak, Y.K. & Song, J. M. (2007). Does the antibacterial activity of silver nanoparticles depend on the shape of the nanoparticle? A study of the gram-negative bacterium *Escherichia coli. Appl. Environl. Microbiol. 73: 1712-1720, ISSN 0099-2240.*

[55] Parameswari, E.; Udayasoorian, C.; Paul Sebastian, S. & Jayabalakrishnan, R.M. 2010. The bactericidal potential of silver nanoparticles. *Intern. Res. J.Biotech. 1(3): 044-049, ISSN 2141-5153.*

[56] Park, E.-S. & Lee, H.-J. (2001). "Synthesis and biocidal activities of polymer. III. Bactericical activity of homopolymer of AcDP and copolymer of acdp with St". *J. Appl. Polym. Sci. 80 (7): 728–736, ISSN 1097-4628.*Proctor, D.L. & Ashman, F. (1972). The control of insects in exported Zambian groundnuts using phosphine and polyethylene lined sacks. *J. Stored Pro. Res. 8: 137-137, ISSN 0022-474X.*

[58] Pyatenko, A.; Shimokowa, K. & Yameguchi, M. (2004). Synthesis of silver nanoparticles by laser ablation in pure water. *J. Appl. Phys. A: Mater. Sci. Proces. A79: 803-806, ISSN 1392-1320.*

[59] Radwan, M.N. & Allin, G.P. (1997). Controlled-release insect repellent device. *US Patent 5,688,509, ISSN 0168-3659.*

[60] Ramsland, T. (J. Selin, Ed.). (1989). Handbook on procurement of packaging. *PRODEC, Toolonkatu 11A, 00100 Helsinki, Finland, ISSN 1456-4491.*

[61] Robertson, G.L. (1993). Food Packaging- principles and practice. *Marcel Dekker, New York, ISSN 0065-2415.*

[62] Scrinis, G. & Lyons, K. (2007). The emerging nano-corporate paradigm: nanotechnology and the transformation of nature, food and agri-food systems. Int. *J. Sociol. Food Agric. 15: pp. 22–44, ISSN 0798-1759.*

[63] Sedlacek, J.D.; Komaravalli, S.R.; Hanley, A.M.; Price, B.D. & Davis P.M. (2001). Life history attributes of indian meal moth (Lepidoptera: Pyralidae) and angoumois grain moth (Lepidoptera: Gelechidae) reared on transgenic corn kernels. *J. Eco. Ento. 94(2): 586-592 ISSN 0022-0493.*

[64] Son, W.K.; Youk, J.H. & Park, W.H. (2006). Antimicrobial cellulose acetate nanofibers containing silver nanoparticles. *Carbohydrate Polymers, 65, 430– 434, ISSN 0144-8617.*

[65] Sondi, I., & Salopek-Sondi, B. (2004). Silver nanoparticles as antimicrobial agent: a case study on *E. coli* as a model for Gram-negative bacteria. *J. Colloid Interf. Sci. 275: 177–182, ISSN 0021-9797.*

[66] Stewart, B. (1995). Packaging as an Effective Marketing Tool. PIRA International, Leatherhead, Surrey, UK, ISSN 1471-5694.

[67] Stout, O.O. (1983). International plant quarantine treatment manual. *Plant Production and Protection Paper, FAO, Rome, 220pp, ISSN 0259-2517.*

[68] Tankhiwale, R., & Bajpai, S.K. (2009). Graft co-polymerization onto cellulose-based filter paper and its further development as silver nanoparticles-loaded antibacterial food packaging material. *Colloid Interf. Sci. 69(2): 164-168, ISSN 0927-7765.*

[69] Taylor, R.W.D. & Harris, A.H. (1994). The fumigation of bag- stacks with phosphine under gas-proof sheets using techniques to avoid development of insect resistance. In: Proceeding of the 6th International Working Conference on Stored- Product Protection, (Edited by: Highley E., Wright E.J., Banks H.J. & Champ B.R.). *CAB International, Wallingford, UK. Vol. 1: 210-214, ISSN 1043-4526.*

[70] Torres, J. A. (1994). Edible films and coatings from proteins. Pages 467- 507 in: Protein Functionality in Food Systems. N. S. Hettiarachchy and G. R. Ziegler, eds. *Marcel Dekker: New York, ISSN 0065-2415.*

[71] Valentini, S.R.T. (1997). Eficiencia de lonas PVC e polietileno para fumigacao de graos comfosfina. *Rev.Bras. de Armaz. Vicosa, 22(1): 3-8, ISSN 1807-1929.*

[72] Watson, E. & Barson, G. (1996). A laboratoryassessment of the behavioural response of *Oryzaephilus surinamensis* (L.) (Coleoptera: Silvanidae) to three insecticides and the insect repellent N, N-diethyl-m-toluamide. *J. Stored Pro. Res. 32: 59–67, ISSN 0022-474X.*

[73] Xie, Y.S.; Fields, P.G. & Isman, M.B. (1995). Repellencyand toxicityof azadirachtin and neem concentrates to three stored-product beetles. *J. Eco. Ento. 88: 1024–1031, ISSN 0022-0493.*

[74] Yu, H.; Xu, X.; Chen, X.; Lu, T.; Zhang, P. & Jing, X. (2007). Preparation and antibacterial effects of PVA–PVP hydrogels containing silver nanoparticles. *J. Appl. Polymer Sci., 103: 125– 133, ISSN 0021-8995.*

[75] Zhang, Y., Chen, F. & Zhuang, J. (2002). Synthesis of silver nanoparticles via electrochemical reduction on Chem. *Commun. (Camb.) 7: 2814-2815, ISSN 1359-7345.*

Shelf Life of Jams in Polypropylene Packaging

Soraia Vilela Borges
Universidade Federal de Lavras,
Departamento de Ciência dos Alimentos,
Brazil

1. Introduction

The shelf life of a product represents the period in which the product remains in good sensory and microbiological for consumption, without harming the taste or health. These conditions are dependent on physical, chemical and microbiological that occur during storage, which depend on the nature of the product, packaging and storage conditions (temperature, relative humidity, storage time) (Man & Jones , 2000).

Jams are generally preserved by applying a combination of obstacles such as lowering the pH, the reduction of water activity by addition of solutes, heat treatment and the use of preservatives and has an expressive consumption in Brazil, given the wide variety existing fruit rich in nutrients, and versatility in the use of these products. From microbiological point of view, these products, according to the packaging and processing conditions and storage, has a shelf life that can vary from 6 months to 1 year (Tfouni & Toledo, 2002), which can be extended by adding sorbic acid and its salts that has good performance in the pH range from 4.0 to 6.0 (Jay 1996).

Among various options for these products packagings stands out the use of polypropylene due to their low water absorption and light (70% at 800nm) compared to cellophane (cheaper), support at high temperatures and low temperatures during the filling and cooling cycle without suffering deformation and be more economical compared to metal packaging (Alves et al., 2007).

Based on the above this chapter aims to report the physical, chemical, physical-chemical, microbiological and sensory occurred in different jams of tropical fruits, stored in jars of polypropylene at different temperature conditions (25-40°C).

2. Materials and methods

2.1 Processing of the jam

The fruits were sanitized with100-ppm chlorine solution, blanched and pulped in a pulper. The jams was processed according to the methodology described by Policarpo *et al.* (2003). The flowchart in Figure 1 contains the elaboration stages of the preserves. Pulp, sugars, calcium carbonate and other ingredients were placed in a stainless steel pan and concentrated to different concentrations according to fruit (72–78°Brix) by heating with constant stirring, according to the formulation. Pectin and acid were added at the end of the cooking period. The preserve was packaged in fixed amounts (100 g) while still hot, using round polypropylene pots (6.7-cm diameter and 7.1-cm height), in cellophane, and molded in these pots, with the same dimensions.

2.2 Experimental design to determine the shelf life

A factorial experimental design: formulations x packaging materials x storage temperature x evaluation times, with two repetitions were used. The physical, chemical, physicochemical and sensorial alterations were evaluated and their means compared by Tukey's test at the 5% level of probability or by statistical models (Cochram & Cox, 1992).

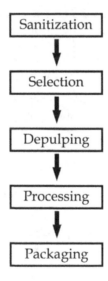

Fig. 1. Flowchart of fruit processing production

2.3 Analytical methods

Chemical and physicochemical analyses were carried out in triplicate according to the following methodologies as described by the Association of Official Analytical Chemists (AOAC 1995): total titratable acidity by titration with NaOH in the presence of phenolphthalein, total and reducing sugars by titration with the Fehling reagent, pH using a potentiometer and total soluble solids by refractometry. Soluble pectin was determined using the methodology described by Bitter and Muir (1962), based on the reaction of the pectin hydrolyzed with carbazol.

Yeasts and molds were determined according to Speak (1976), making serial dilutions with peptone water and plating on potato dextrose agar with or without 18% glucose to detect osmophilic and other species of yeast. Plates were incubated at 25–30 C, and the results were analyzed after 48 and 72 h of incubation and expressed in colony-forming units per gram. Water activity (Aw) was measured at 33 °C in an AquaLab device (model Cx-2, Decagon Device, Pullman, WA).

Color parameters (L*, a*, b* and hue angle) were determined in a Minolta colorimeter (model CM-3600d, Konica Minolta,Ramsey, NJ), using the standard D65/10° illuminant, without including the shine factor.

Texture profile analysis (TPA) was determined using the TA.TX2i texturometer (Stable Micro Systems, Surrey, UK), operated with the Texture Expert software, using a pretest velocity of 2 mm/s, test velocity of 1 mm/s and posttest velocity of 2 mm/s, with a distance of 5 mm and time of 5 s, the test body being acrylic (P25/L) with a diameter of 25 mm. The product was placed in an aluminium capsule with a diameter of 50 mm and height of 25 mm for standardization. A graph of force x time was constructed, each characteristic peak corresponding to one of the TPA characteristics.

The sensory evaluation of the jams was carried out by 60 potential consumers and the attributes appraised (appearance, aroma, texture, color, flavor and global impression) judged using a structured 9 point hedonic scale (1= disliked extremely to 9= liked extremely). The experiment was applied to a balanced complete block design according to Stone and Sidel (2004), and 20 g of each treatment presented in plastic cups codified with three digit numbers. The tests were carried out in individual booths in the food sensory analysis laboratory. The results were submitted to an analysis of variance (ANOVA) and the differences between the averages compared by the test of Tukey at 5% of probability (Cochram & Cox, 1992).

3. Results

In several studies on the shelf life to tropical fruit jams (banana, guava, umbu, shell passion fruit juice) packaged in polypropylene, physical, physicochemical, chemical and sensory characteristics were recorded during storage, while microbiological characteristics these products were stable for 4-6 months of storage at temperatures of 25-40 ° C (Policarpo et al. 2007; Nascimento, 2002, Menezes et al. 2011; Martins et al, 2010; Martins et al, 2011). This is due to the fact they were manufactured using good manufacturing practices, the products are low pH and high concentration of sugar, and good sealing of packaging, conditions that minimize or prevent microbial growth of fungi and yeasts, this typical range of water activity.

Regarding the physicochemical properties, there was a slight drop in pH and concomitant increase in titrable acidity (Martins, 2009, Martins et al. 2010, Nascimento et al., 2002) and in some products the pH remained unchanged (Policarpo et al.,2007, Menezes et al. 2011). Soluble solids tend to increase due to syneresis or water evaporation during storage at high temperatures as observed in the work of Policarpo et al. 2007; Menezes et al. 2011; Martins et al., 2010; Martins et al. 2011. The syneresis due to conditions of low pH of the gel formed and inability to retain moisture in the product and considering that polypropylene has a certain permeability is possible the migration of water into the environment during storage, especially the high temperatures (Fizman & Duran, 1992). The total sugars tend to increase for the same reason the total solids (effect of concentration) and sucrose hydrolysis occurs in glucose due to the increased acidity (Policarpo et al., 2007, Menezes, et al. 2011, Martins et al., 2010). These sugars are hygroscopic and reduce the water activity during storage (Martins et al., 2011). Pectins, when added to preserves or jelly to stabilize the gel network formed and increase the firmness, are also hydrolyzed by the increase in acidity during storage, especially at higher temperatures (Policarpo et al., 2007).

Color parameters (L *, a *, b *) and texture profile, depending on the chemical changes that occur are also altered. Was noticed a decrease in the value of L * in all the above products, indicating darkening, which is due to several reactions accelerated by high temperatures, and exposure to light, such as oxidation of vitamins and pigments in fruits and others reactions. In parallel b parameters relating to color tone characteristic of the fruit, by factors similar to the luminosity are reduced (Policarpo et al. 2007; Martins et al, 2010) or increased (Menezes, et al.,2011; Martins et al., 2011), according to the processed fruit. In the texture profile in all work we observed an increase in firmness with increasing soluble solids, and in some cases increasing of gumminess, due to syneresis. Other parameters of texture profile analysis showed no significant changes.

Regarding the sensory evaluation tests for affective attributes of color, texture, flavor and overall impression, using a hedonic scale of 9 points and untrained judges (Stone & Sidel, 2004) showed that there is reduction in all attributes during the storage. For jelly albedo of passion fruit / passion fruit juice the result of global acceptance at 90 days/25° C were averaging close to 7, corresponding to liked moderately (Nascimento et al., 2002). For guava preserve showed an average between 7-6 (like slightly-like moderately) for different attributes to 150 days of storage at 20 ° C, and detected the appearance of crystals after 90 days, enhanced by the use of potassium sorbate as a preservative (Menezes, 2008). For banana cv prata preserve, kept at 20 -30 ° C for 75 days had a good overall acceptance (in the scores in the range 6-7).

4. Conclusions

The packaging of polypropylene due to low gas permeability and light is appropriate and economical for a short time to market for packaging fruit jams (up to 150 days), maintaining acceptable products in terms of sensory and microbiological changes.

5. Acknowledgements

At FAPERJ (Rio de Janeiro) and FAPEMIG (Minas Gerais) for financial support to these projects, CNPq and CAPES for the scholarships awarded to students of post graduate and undergraduate; to EMBRAPA (CPATSA-Petrolina, CTAA, Rio de Janeiro-RJ) SENAI (CETIQT-Rio de Janeiro, CETEC-Vassouras) for collaborating in different analysis of these projects, students, teachers, researchers and contributors of UFRRJ, UFLA, EMBRAPA and ITAL.

6. References

Alves, R.M.; Dender, A.G.F. Van; Jaime, S.B.M.; Moreno, I. & Pereira, B.C. (2007).Effect of light and packages on stability of spreadable processed cheese. *International of Dairy Journal*, Vol. 17, No. 4, pp. 365-373, ISSN 09586946

Association of Official Analytical Chemists (AOAC) (1995) *Official Methods of Analysis of the Association of Analytical Chemistry*. AOAC, ISBN 0935584544, Arlington.

Bitter, T. & Muir, H.M. (1962). A modified uronic acid carbazole reaction. Anal. Biochem. V.4, No. 4, pp. 30–334. ISSN 003-2697

Cochram G. & COX, G.M. (1992) *Experimental Design*, J. Willey, ISBN 0477545678, New York.

Fizman, S.M. & Duran, L. (1992). Effect of fruit pulps and sucrose on the compression response of different polysaccharides gel systems. *Carbohydrate Polymers*, Vol.17, No.1, pp.11-17, ISSN :01448617

Jay, J.M. (1996). *Modern Food Microbiology*, Chapmam & Hall, ISBN 0442244452, New York.

Man, C.M.D. & Jones, A.A. (2000). *Shelf life Evaluation of Foods*. IBlackie Academic and Professional, ISBN 0834217821, Glasgow.

Martins, M.L.A.; Borges, S.V.; Cunha, A.C.; Oliveira, F.P.; Augusta, I.M. & Amorim, E. (2010). Alterações físico-quimicas e microbiológicas durante o armazenamento de doces de umbu (Spondias tuberosa Arr. Câmara) verde e maduro. *Ciência e Tecnologia de Alimentos*, Vol.30, No.1., pp. 6-7, ISSN 0101-2061

Martins, G.S.M.; Ferrua, F.Q.; Mesquita, K.S.; Borges, S.V.; Carneiro, J.D.S. (2011). Estabilidade de doces em massa de banana prata. *Revista do Instituto Adolfo Lutz*, Vol. 70, No. 1, pp. 1-20, ISSN 0073-9855.

Menezes, C.C.; Borges, S. V.; Ferrua, F.Q., Vilela, C.P.; Carneiro, J.D.S. (2011). Influence of packaging and potassium sorbate on the physical, physicochemical and microbiological alterations of guava preserves. *Ciência e Tecnologia de Alimentos*, Vol.31, No.3., pp. 674-680, ISSN 0101-2061.

Menezes, C.C. (2008) *Otimização do processo de elaboração e avaliação da presença de sorbato e embalagens sobre o doce de goiaba durante armazenamento*. UFLA, Lavras.

Nascimento, M.R.F.; Oliveira, L.F. & Borges, S.B. Estudo da conservação de doce de corte de casca do maracujá à temperatura ambiente. *Proceeding of the 13th Congresso Brasileiro de Ciência e Tecnologia de Alimentos*. August, pp. 4-7, 2002, Porto Alegre, Brazil (cd rom)

Policarpo, V.M.N.; Borges, S.V.; Endo, E.; Castro, F.T.; Anjos, V.D. & Cavalcanti, N.B. (2007). Green umbu (Spondias Tuberosa Arr. Cam.) preserve: physical, chemical and microbiological changes during storage. *Journal of Food Processing and Preservation*, Vol. 31, No. 2, pp. 201-210, ISSN 01458892

Tfouni, S.A.V. & Toledo, M.C.F. (2002). Determination of benzoic and sorbic acids in Brasilian food. *Food Control*, Vol.13, No.2, pp. 117-123, ISSN 09567135

Speak, M.L. (1976) *Compendium of Methods for the Microbiological Examinations of Foods*. ISBN 0875530818, American Public Health Association, Washington.

Stone, H. & Sidel, J. (2004). *Sensory evaluation practices*. ISBN 0126726906, Academic Press, New York.

3

Thermal Oxidation of Polypropylene and Modified Polypropylene – Structure Effects

Lyudmila Shibryaeva
N.M. Emanuel Institute of Biochemical Physics,
Russian Academy of Sciences,
Russia

1. Introduction

Ageing and stabilization polymers is a major part of materials science. Aging of polymers defined as a set of chemical and physical transformations, leads to the loss of their set the desired properties. The main role in these transformations belong to chemical processes of degradation and crosslinking of macromolecules. Processes decomposition and structuring polymer conjugates include radical - chain, ionic and molecular reactions. Traditionally, the distinction is made between thermal, thermal-, photo-and radiation-chemical aging. Thermal oxidation and thermal-oxidative destruction are the most common and important processes in which polymer materials participate. It accompanies the posed for chemists-experimenters and producers engaged in polymer materials creation is the problem of maintenance of high quality of output, prolongation of its service life at conditions of thermal oxidation influence and thermal-oxidative destruction. Thermal oxidation of polymers leads to a modification and functionalization of the polymer chains. At the same time thermal oxidation is accompanied by the destruction of bonds in the macromolecules and influence the destructive processes. Thermal oxidation of polymers - a radical chain process with degenerate branching of kinetic chains of oxidation. The structure of polymer significantly influences on chain oxidation and destructive processes. Heterogeneity of polymers structures, the presence of regions differing in amplitudes of molecular motions, decrease of segment mobility, reduction of oxygen diffusion coefficient underlie this effect. These factors change kinetics and mechanism of process. As new methods and polymeric materials, researchers returned to the discussion of the induction period of oxidation of polymers. However, often in the literature there is confusion in the very concept and definition of the period induction. For example, when studying the thermal oxidative degradation of PP with different tacticity by thermogravimetric analysis (Chan J.H., Balke S.T., 1997; Nakatani H. and al., 2005) determine the induction period as a time corresponding to the onset of weight loss. I.e in fact, the period induction regarded not as the beginning of oxidation in infancy kinetic chain, and the time corresponding to the branching of kinetic chains in the collapse of hydroperoxide. This is a fundamental difference, so it is important this issue be considered. Often thermal-oxidative degradation is identified with the thermal oxidation. However, in depending on the nature of the polymer, thermal oxidation process can take place without destruction chains, and with functionalization. As in the case of isotactic polypropylene. The author of the chapters of the book sets the task of separating the concepts related to the kinetics and mechanism.

2. The particularities of oxidation of polypropylene

2.1 The kinetics and mechanism of autooxidation of solid polymers

Three types reactions can be in the polymers in the presence of oxygen. 1) Separately occurring molecule reactions. 2) The radical - chain mechanism 3) The products of thermal decomposition and oxidation of polymers catalyze further decomposition of the polymer. The thermal oxidation of polyolefynes has been extensively investigated in various works. The investigation of the kinetics and mechanism of oxidation of solid polymers have shown convincingly that this process is a radical - chain with degenerate branching of kinetic chains. In the thermal degradation, thermooxidation and thermal oxidative degradation of polymers play a major role alkyd (R^*), alkoxide (RO^*) and peroxide (RO_2^*) macroradicals and low molecular weight radicals (r^*). The high reactivity of the past towards macromolecules strongly influences on aging processes. The chain reaction of the oxidation of a polymer includes alternate steps of the chain propagation proceeding either inside the same macromolecule or between two molecules. The investigation of kinetics of oxidation of the polymers, containing aliphatic groups ($\equiv$ C-H, -CH_2- or –CH_3), showed that this process was described by scheme, corresponding to the mechanism of chain oxidation of liquid phase (Denisov E.T. and all., 1975).

2.1.1 Initiation of kinetic chain of oxidation

For the oxidation of the polymer to the formation of macroradicals R *.

$$RH \longrightarrow R^* + H^* \tag{1}$$

Where RH - the monomer units of polymer. Reaction can be triggered by physical factors such as ultraviolet and ionizing radiation, heat, ultrasound, or mechanical treatment chemical factors, such as catalysis, a direct reaction with molecular, singlet or atomic oxygen and ozone. However, initiation by direct interaction of molecular oxygen with the polymer, leads to detachment of a hydrogen atom, was unlikely, because it is endothermic reaction, enthalpy is 126-189 kJ/mol (Chan J.H., Balke S.T., 1997). Often, the birth of the chain portrayed as the bimolecular interaction of oxygen with the monomer units of polymer

$$RH + O_2 \longrightarrow [RHO_2] \longrightarrow R^* + HO_2^* \tag{2}$$

HO_2^* radicals, which formed, can enter on reaction with neighboring RH or on reaction of recombination with the primary radical R *

$$RH + O_2 \xrightarrow{k_o} [R^* + HO_2^*;RH] \longrightarrow \sigma R^* \tag{3}$$

Therefore, the radical yield (f) is: 0 <σ<2. RH- may be neighboring monomer units of one macromolecule or belong to different macromolecules. At the origin of the chain oxidation may participate impurities of transition metals, residues of catalysts or initiators, etc. These impurities get into the polymer as a result of receiving or processing the polymer. Table. 1

shows the rate of nucleation of chains, obtained by different authors inhibitor method, calculated from the rate of inhibitor consumption in polyethylene, polypropylene and some liquid hydrocarbons. The values of W_o are small, so usually, when considering the kinetics of the autoxidation of polyolefins, nucleation rate of the chain is neglected compared with the rate of branching.

Substance	T,K	p_{O_2} millimeters of mercury	W_{O_2}, mol/l• s	E, kJ/mol
PEHD	404	750	$1.3 \cdot 10^{-6}$	117.0
PELD	377	750	$3.4 \cdot 10^{-7}$	146.5
PE	391	750	$1.5 \cdot 10^{-7}$	-
PE melt	473	300	$8.0 \cdot 10^{-7}$	-
PP	405	750	$2.1 \cdot 10^{-6}$	92.0
	403	750	$2.4 \cdot 10^{-6}$	-
Atactic PP	403	300	$<1.2 \cdot 10^{-8}$	-
	423	300	$<6.0 \cdot 10^{-8}$	-
Isotactic PP	453	300	$< 7.0 \cdot 10^{-8}$	-
	463	300	$< 2.7 \cdot 10^{-6}$	-
2-Methylbutane	473	300	$< 5.6 \cdot 10^{-6}$	-
	410	4725	$2.2 \cdot 10^{-9}$	159.0
n-Heptane	406	4100	$1.4 \cdot 10^{-9}$	181.0

Table 1. The kinetic parameters of nucleation reaction chain (RH +O2) in polyethylene, 7 polypropylene and liquid hydrocarbons

2.1.2 Growth of the chain

The development of the kinetic chain by alternation of two reactions: the formation of peroxide radicals (RO$_2$ *) and hydroperoxide (ROOH). Macroradicals R*, appeared in the initiation can easily react with oxygen molecules to give peroxide radicals RO$_2$ * Peroxide radical can pull hydrogen from another polymer molecules to form polymeric hydroperoxides:

$$R * + O_2 \xrightarrow{k_1} RO_2 * \tag{4}$$

$$RO_2* + RH \xrightarrow{k_2} \alpha ROOH + R* \tag{5}$$

where k_2 - the constant of continuation of kinetic chain rate.α - the yield of hydroperoxide per mole of absorbed oxygen. In the solid polymer free radical R * and hydroperoxide group, formed in reaction (5) can not be away from each other. Part of ROOH is destroyed immediately after the formation of the reaction:

$$ROOH + R* \longrightarrow RO* + ROH \tag{6}$$

The reaction of (6) leads to a decrease in the yield of hydroperoxides during the oxidation of polymers in comparison with the oxidation of liquid hydrocarbon model. Their output ROOH is close to 100%. In the presence of oxygen even at low concentrations of the radicals

R* are converted into RO_2^* continue to ROOH. The concentration of the radicals R* is negligible compared to RO_2^*, so oxidation rate (Wo_2) is determined (limited) reaction rate (5). In this case: $Wo_2 = k_1 [R^*] [O_2] = k_2 [RO_2^*] [RH]$.

2.1.3 The stage of branching of kinetic chain

Branching of the kinetic chain of oxidation occurs in the decay of polymer hydroperoxides. Generally, consider a few basic mechanisms of decomposition of hydroperoxide

$$ROOH \xrightarrow{k'_d} RO^* + OH^* \qquad (7)$$

$$ROOH + RH \xrightarrow{k_d} RO^* + R^* + H_2O \qquad (8)$$

$$2\,ROOH \xrightarrow{k''_d} RO^* + H_2O + RO_2^* \qquad (9)$$

where k_d, k'_d, k''_d - the constants of ROOH decomposition rate.

Monomolecular decay (7) comes with a large activation energy (140-160 kJ / mol). It occurs only in the oxidation of hydrocarbon fluids in the case of low concentrations of ROOH in solvents not containing weakly bound hydrogen atoms. Are more favorable reaction (8) and (9). Heat of reaction (9) is ~ 36 kJ / mol, and for reaction (8) varies widely depending on the binding energy of the R-H. Reaction (9) dominates at high concentrations of hydroperoxide, the reaction (8) - in small quantities. In polymers containing weakly bound hydrogen atoms are predominant mechanism (8). As usual [ROOH] <<[RH], ROOH decay is described by a kinetic equation of first order.

2.1.4 Chain transfer

The particularity of harden phase oxidation of polyolefyne is reaction of chain transfer – interaction of alkyl (R*) or alkoxy radical (RO*) with polymer competitive to its reaction with oxygen:

$$R^* + R'H \xrightarrow{k''_2} R'^* + RH \qquad (10)$$

$$[R'^*, R''^*, RO^*, OH^* + RH \longrightarrow R^* + R'H, R''H, ROH, H_2O]$$

2.1.5 Chain termination

Break radical chain due to the interaction of free radicals with each other to form inactive products. There is quadratic termination of peroxide radicals at high pressure of oxygen:

$$RO_2^* + RO_2^* \xrightarrow{k_t} O_2 + molecular\ products \qquad (11)$$

Chain termination at low pressure of oxygen is quadratic termination of alkyl radicals

$$R^* + R^* \xrightarrow{k_4} R - R \qquad (12)$$

and alkyl with peroxide radicals:

$$RO_2^* + R^* \xrightarrow{k_5} ROOR \tag{13}$$

Where k_t, k_4, k_5 – the constants of chain termination rate.

2.2 The kinetics and mechanism of autooxidation PP

Oxidation PP occurs in the amorphous regions of the polymer. Localization process in the amorphous regions was confirmed by small-angle X-ray scattering, by direct measurements of oxygen solubility in the samples with varying degrees of crystallinity, by the spin-paramagnetic resonance and other methods. The soluble oxygen, impurities, that contribute to the initiation of oxidation (traces of polymerization catalysts, traces of carbonyl groups, hydroperoxide and unsaturated groups) are localized in the amorphous regions polymer. This leads to a higher initiation rate in the amorphous areas compared to the total weight of the polymer. In the crystalline phase of PP on steric reasons, prohibited further kinetic chain reaction of oxidation. Even with the presence of peroxide radicals in the crystalline phase are not involved in the development of kinetic chains of oxidation of the crystallites. These radicals can formed by the action of γ- radiation on the polymer. RO_2^* slowly dying in the crystallites by the decay of education low-molecular radicals, which may go into an amorphous phase, initiating there oxidation.

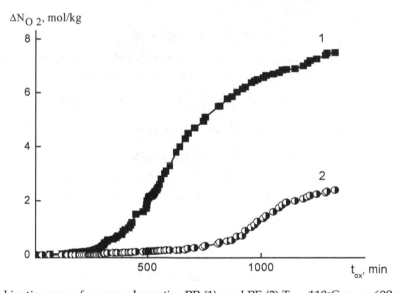

Fig. 1. The kinetic curve of oxygen absorption PP (1), and PE (2) T_{ox} =110°C, p_{o2} = 600 mm Hg

Oxidation of polypropylene describes the kinetic curve of oxygen absorption, which has an S-shape (fig.1). This curve is characterized by an induction period of self-acceleration and deceleration of oxidation in a deep stage of the process. A typical kinetic curve of oxygen uptake for isotactic PP is shown in Figure 1. For comparison, the kinetic curve for polyethylene (PEHD). The kinetic equation represents the dependence of the amount of absorbed oxygen from the oxidation time was determined applied to the polypropylene in (Emanuel N.M., Buchachenko A.L., 1982). With this purpose it was used a special mathematical model for handling the above proposed scheme of oxidation of solid polymers. It was suggested the following approach: the origin of kinetic chain occurs at hydroperoxide decomposition (equation (8).The rate of this reaction (which is reaction of degenerate branching of kinetic chains) in the early stages process considerably exceeds the rate of primary initiation (1).This allows to neglect the primary initiation reaction (1) and assume that the nucleation rate kinetic chain is the speed of degenerate branching. In this case, the initiation rate is determined from the equation:

$$w_i = 2\delta k_d[RH][ROOH]$$

where δ- probability going out of the radicals on one hydroperoxide group which is broken, i.e. probability of degenerate branching, 2δ - number of kinetic chains, born of every molecule of hydroperoxide decomposed. Type of reaction, limiting the development of kinetic chains (4) or (5), depends on the oxygen concentration $[O_2]$, consequently on the concentration of radicals R^* and RO_2^*. Concentration ratio $[R^*]/[RO_2^*]$ from the condition

$$k_1[O_2][R^*] = k_2[RH][RO_2^*]$$

whence $[R^*]/[RO_2^*] = k_2[RH]/k_1[O_2] \cong 10^{-2} \div 10^{-4}$. Thus, the kinetic chain termination occurs only with peroxide radicals. The termination of kinetic chain occurs by a quadratic law:

$$2RO_2^* \xrightarrow{k_t} ROOR + O_2$$

Kinetic equations for determining the rate of oxygen absorption can be obtained from the kinetic equations for the intermediate concentrations of particulate matter:

$$d[R^*]/dt = 2\delta k_d[RH][ROOH] + k_2[RH][RO_2^*] - k_1[O_2][R^*] \qquad (14)$$

$$d[RO_2^*]/dt = k_1[O_2][R^*] - k_2[RH][RO_2^*] - k_t[RO_2^*]^2 \qquad (15)$$

$$d[ROOH]/dt = k_2[RH][RO_2^*] - k_d[RH][ROOH] \qquad (16)$$

Since the slow processes of oxidation can be applied to the quasi-steady concentration of radicals, i.e. $d[R^*]/dt = d[RO_2^*]/dt = 0$.From the equation (14) and (15) follows the expression

$$[RO_2^*] = (2\delta k_d[RH][ROOH]/k_t)^{1/2} \qquad (17)$$

Where the rate of oxygen uptake

$$d\Delta[O_2])/dt = k_1[O_2][R^*] = 2\delta k_d[RH][ROOH] + k_2[RH][RO_2^*] =$$

$$= 2\delta k_d[RH][ROOH] + k_2(2\delta k_d/k_t)^{1/2}[RH]^{3/2}[ROOH]^{1/2} \qquad (18)$$

$\Delta[O_2]$ - means the amount of oxygen, which absorbed at a given moment of time.

From equations (14) and (16)

$$d(\Delta[O_2])/dt = d[ROOH]/dt + (1+2\delta)k_d[RH][ROOH] \qquad (19)$$

From equation (19) that since the maximum concentration of hydroperoxide ($d[ROOH]/dt = 0$), the rate of oxidation is proportional to the speed of its disintegration. The proportionality factor ($(1+2\delta)$ can vary from 1 to 3 (as $0 \leq \delta \leq 1$). Substitute equation (17) in the equation (16) gives:

$$d[ROOH]/dt = k_2(2\delta k_d/k_t)^{1/2}[RH]^{3/2}[ROOH]^{1/2} - k_2[RH][ROOH] \tag{20}$$

Integration (20) at $[RH]$ = const, gives the equation:

$$[ROOH] = [ROOH]_{max}\{1-[1-([ROOH]_o/[ROOH]_{max})^{1/2}]exp(-k_d[RH]\ t/\ 2)\}^2 \tag{21}$$

Where $[ROOH]_{max} = (2\delta k_2^2/\ k_d\ k_t)\ [RH]\ t$; $[ROOH]_{max}$, $[ROOH]_o$ -maximum and the initial concentration of hydroperoxide, respectively. So far as $[ROOH]_o \ll [ROOH]_{max}$,value ratio $[ROOH]_o/[ROOH]_{max}$ - infinitely small quantity, which can be neglected. In this case the expression (21) takes the form:

$$[ROOH] = [ROOH]_{max}[1-exp((-k_d[RH]/2)t)]^2 \tag{22}$$

Substituting equation (19) in (15) and integrating gives the expression

$$\Delta[O_2]/[ROOH]_{max} = k_d(1+2\delta)[RH]t + 2(1+4\delta)[exp(-k_d[RH]t/2)-1]-$$

$$-2\delta[exp(-k_d[RH]t)-1] \tag{23}$$

Equation (23) represents the integral form of the dependence of the amount of absorbed oxygen from the oxidation time. This dependence is valid for shallow oxidation, we can neglect the flow of the polymer and assume $[RH]$ = const. Expanding the exponential terms in the series for small t, and only the first three terms, gives the expression:

$$\Delta[O_2] = (k_2^2\delta k_d[RH]^3/\ 2k_t)t^2 = \Phi^2t^2 = 1\ /\ 2\ [ROOH]_{max}k^2_d[RH]^2\ t^2 \tag{24}$$

So. oxygen uptake at the beginning of oxidation should be proportional to t^2; Φ - self-acceleration factor of the reaction.

$$\Phi = (1\ /.2^{1/2})\ (k_2/k^{1/2}_t)\ (\delta k_d)^{1/2}[RH]^{3/2}\ or \quad \Phi = (1\ /.2^{1/2})[ROOH]^{1/2}_{max}\ k_d[RH] \tag{25}$$

Substituting this expression in (20) and assuming that the oxidized monomer units of the polymer in the subsequent oxidation do not participate, gives:

$$d[ROOH]/dt = k_2(2\delta k_d/\ k_t)^{1/2}([RH]_o-\Delta[O_2])^{3/2}[ROOH]^{1/2} - k_d([RH]_o-\Delta[O_2])[ROOH \tag{26}$$

$$d(\Delta[O_2])/dt = k_2(2\delta k_d/k_t)^{1/2}([RH]_o-\Delta[O_2])^{3/2}[ROOH]^{1/2} - 2\delta k_d([RH]_o-\Delta[O_2])[ROOH] \tag{27}$$

From some point set a quasi-stationary concentration of hydroperoxide,

i.e $d[ROOH]/\ dt = 0$. Under this condition, equation (26) can be simplified

$$[ROOH] = k_2^2\delta\ /\ k_d\ k_t = ([RH]_o - \Delta[O_2]) \tag{28}$$

i.e quasistationary concentration of hydroperoxide is proportional to the concentration of unoxidized polymer. When substituting this relation into equation (27) can be obtained

$$d(\Delta[O_2])/dt = 2 (1+2\delta) \delta k^2_2/k_t ([RH]_o - \Delta[O_2]) \tag{29}$$

After integrating the initial condition $\Delta[O_2]= (\Delta[O_2])_1$ and $t=t_1$, is obtained

$$1/ ([RH]_o-\Delta[O_2]) = 2 (1+2\delta)(\delta k^2_2/k_t)t_1 +\alpha \tag{30}$$

whence

$$\alpha = 1/ ([RH]_o-\Delta[O_2]) - 2 (1+2\delta)(\delta k^2_2/k_t)t_1, \tag{31}$$

where t_1 means the time from which the concentration of hydroperoxide can be considered quasi-stationary. , it is still time to reach maximum concentration of hydroperoxide. Another model, describing the kinetics oxidation of PP, based the same pattern as discussed in (Shlyapnikov Yu. and al., 1986). This work identified two stages of the oxidation process. The initial stage of reaction and phase deep oxidation. At the initial stage of the reaction rate of hydroperoxide significantly higher than the rate of its thermal decomposition. The latter can be neglected. The equation of balance of free radicals and hydroperoxide in amorphous material in the absence of a linear chain termination is as follows:

$$d[RO_2{}^*]/dt = W_0 + \alpha k_d[RH][ROOH]- 2k_t [RO_2{}^*] \tag{32}$$

$$d[ROOH]/dt = \alpha k_2[RH][RO_2{}^*] \tag{33}$$

System of equations (32) and (33) is solved in a quasistationary approximation $d[RO_2{}^*]/dt=o$; $d[ROOH]/dt\neq0$ The solution of equation (31) has the form:

$$[RO_2{}^*]= \{W_o + \sigma k_d[RH][ROOH]/2k_t\}^{1/2}, \tag{34}$$

where $W_0 = f k_o [RH] [O_2]$ - rate of nucleation of chain , which negligible compared with that of chain branching in the early stages process. Substituting the values of $[RO_2 {}^*]$ (34) and (33) and integration of this expression gives the variation of the expression ROOH concentration in time:

$$[ROOH] = \alpha N_{o2}= \alpha \cdot 2\sigma k_2{}^2k_d[RH]^3t^2/8k_t = At^2 \tag{35}$$

So manner of equation (35) shows that in the initial stage reaction oxidation of the amount of oxygen absorbed during oxidation, and concentration hydroperoxide proportional to the square of oxidation time. This is consistent with the conclusion drawn in previous work (Emanuel N.M., Buchachenko A.L.,1982). Comparison (36) and (37) provided in the form of dependence changes in the concentration oxygen from the oxidation time:

$$\Delta[O_2] = (k^2_2\delta k_d[RH]^3/ 2k_t) t^2 = \Phi^2t^2 \tag{36}$$

$$N_{o2} = (\sigma k_2{}^2k_d[RH]^3/4k_t) t^2= At^2 \tag{37}$$

show that the process oxygen uptake in the initial phase described by a parabolic law. For PP parabolic law is obeyed up to $\Delta[O_2]= 1$ mole/kg. At a more profound stage of a deviation from this law. Deviations from parabolic law is also observed at short times of oxidation, when still not satisfied quasistationarity the concentration of hydroperoxide (in the induction period).

2.3 Oxidation polymer on the deep stage

At deep stages of oxidation the rate of decomposition of hydroperoxide increases with its concentration. The rate of quadratic chain termination is proportional to the square concentration of radicals. At this point plays an important role expenditure monomer units of the polymer. Nucleation rate of the chain compared with the rate of branching can be neglected. The reaction is carried out at a high concentration of oxygen and the contribution reaction $R^*+RO_2^*$ is negligible, $[R^*]<<[RO_2^*]$. For these conditions, the balance equation of free radicals and hydroperoxide is:

$$d[R^*]/dt = k_2[RH][RO_2^*] - k_1[O_2][R^*] + \sigma k_d[RH][ROOH] \tag{38}$$

$$d[RO_2^*]/dt = k_1[O_2][R^*] - k_2[RH][RO_2^*] - 2k_t[RO_2^*]^2 \tag{39}$$

$$d[ROOH]/dt = \alpha k_2[RH][RO_2^*] - k_d[RH][ROOH] \tag{40}$$

Using the method of quasistationary concentrations for $[RO_2^*]$ can be obtained:

$$[RO_2^*] = (\sigma k_d[RH]/2k_t)^{1/2}[ROOH]^{1/2} \tag{41}$$

$$d[ROOH]/dt = \alpha\sigma^{1/2}\{k_2k_d^{1/2}[RH]^{3/2}/\sqrt{2}k_t^{1/2}\}[ROOH] - k_d[RH][ROOH] \tag{42}$$

The solution of equation (40) $[ROOH]o = 0$ has the form:

$$[ROOH] = [ROOH]_{max}[1 - \exp((-k_d[RH])/2) t)]^2 \tag{43}$$

$$[ROOH]_{max} = \alpha^2\sigma k_2^2[RH]/2k_dk_t \tag{44}$$

$[ROOH]_{max}$ –limit sought by the concentration of hydroperoxide in the case of $[RH] = const.$

In the same conditions, the rate of oxygen consumption varies as

$$Wo_2 = dNo_2/dt = \alpha\sigma k_2^2[RH]^2/ k_dk_t [1 - \exp((-k_d[RH])/2) t)] \tag{45}$$

Equations (23), (24) describe the dependence of the amount of absorbed oxygen from the oxidation time and the equation takes into account the flow of the polymer during the reaction (31) allow us to establish the oxidation characteristics of polypropylene and the factors influencing this process. The above equation (35) does not take into account the flow of the polymer during the reaction. Because the rate of consumption of the polymer is the rate of absorption of oxygen:

$$d[RH]/dt = d(\Delta[O_2])/dt, \tag{46}$$

$$[RH] = [RH]_o - \Delta[O_2], \tag{47}$$

Where $[RH]_o$, $[RH]$ - the concentration of polymer in the beginning of the reaction and at time t; $\Delta[O_2]$ – the amount of oxygen, absorbed by this time.

2.4 The reaction to continue the kinetic chain of oxidation $RO_2^* + RH \longrightarrow ROOH + R^*$

A key step in the radical chain oxidation of polypropylene is the reaction to continue the kinetic chain of oxidation (5). It determines the overall speed of the process leads to spatial displacement of the free valence, it is the product of hydroperoxide, which is the degenerate branching of kinetic oxidation chains. In polyolefins, the reaction can take place both inside and intermolecularly. The oxidate polymer chain fragment may be a part of one macromolecule to which the RO_2^* radical belongs or a part of the neighbouring macromolecule. The consequences of the intra- and intermolecular chain propagation are different for . various polyolefins. So, in polyethylene, the main mode of reaction is intermolecular (Emanuel N.M., Buchachenko A.L., 1982). In contrast to PE, in polypropylene, this reaction is carried out intramolecularly (Shlyapnikov Yu. A. and al., 1986; Rapoport N.Ya. and al, 1986). The reason for the differences between the PP and PE is the conformational structure of the polymer chain. With the continuation of the intramolecular kinetic chain the probability of activated complex optimal structure should depend on the local conformation of the site of a macromolecule carrying a peroxide macroradical. Conformation of the reaction complex is given below:

$$
\begin{array}{c}
H \\
\backslash \qquad\qquad H \\
H{-}C{\cdots}H{\cdots}O \qquad / \\
/ \qquad\qquad \backslash O{-}C{-}H \\
H \qquad\qquad\qquad \backslash \\
\qquad\qquad\qquad\qquad H
\end{array}
$$

In this complex bond angles $\angle$H-C-H) = 109°, $\angle$C-H-O) = 180°, $\angle$HOO = 100-105°, the distance between the atoms O···H = 1,4 Å , C·H =1,2 Å. Reaction does not occur, if the distance O···H more than 1,8 Å (Rapoport N.Ya., Mostovaya E.M., and all 1986). With the help of analysis of molecular models of Stuart-Briegleb, it was show that the probability of formation of activated complex of optimal structure depends on the set of conformations of the macromolecule. For example in PE, the linear activated complex is not formed in a macromolecule, having a straightened conformation of trans - zigzag, consisting of a sequence of trans-conformers (~ TT ~). For the occurrence of an intramolecular reaction of the chain oxidation of PE required sequence of two folded gosh-conformers type GG or G·G· . Where are the angles of internal rotation around C-C bonds for G and G-conformers are equal to 120°, the angles of internal rotation for T-conformers are equal 0°. However, the equilibrium fraction of dyads GG in PE low, at room temperature it is approximately 9%. On the other hand, in PE not bulky lateral substituents create steric hindrance to intermolecular continuation of the kinetic chain of oxidation. Due to these factors, education linear activated complex with PE in intermolecular reaction is realized inorder are more likely than in intramolecular (Popov A.A. and al. 1987). The presence of side substituents in the macromolecule of polypropylene leads to the formation of helical conformation, which consists of a series of trans and gauche-conformers ~ TGTGTG ~. This sequence forms an extremely straightened conformation of the chain, which is "hard", and corresponds to the minimum energy on potential curve of interaction of valence-not bonded atoms. Folded conformation is formed by alternating joints left-and right dextro-rotatory sites spirals type ~TGTGTGG*TGT~. Where G* - gauche conformer with the angle of internal rotation, deviating from 120°, relevant ± 60°. The relative position of atoms H adjacent tertiary C-H bonds in PP is determined by the type of dyad conformer: TT, TG and GG. Dyad TT in PP is not implemented due to the overlap of methyl groups in dyad TG peroxide radical is shielded from neighboring tertiary connection CH. Education of the reaction complex is most likely if the conformation of the reaction center meets the dyad GG *. This dyad is a prelaunch conformers for the intramolecular reaction. This explains why in contrast to PE in polypropylene the reaction of RO_2 * + RH proceeds mainly intramolecularly. In (Rapoport N.Ya. and al, 1986), this reaction was considered from the perspective of a reaction "pair" the radical RO_2 * - GG *-conformer:

$$RO_2^* + GG^* \overset{k_r}{\underset{k_{-r}}{\Leftrightarrow}} [RO^*_2 (GG^*)] \overset{k_x}{\longrightarrow} ROOH$$

For the case $k_r \simeq k_{-r}$ speed of reaction in the quasistationary approximation is:

$$W_2 = (k_x k_r / k_x + k_r)[RO^*_2][GG^*] = k_{3\phi}[RO^*_2][GG^*]$$

If the limiting stage will be meeting with conformer GG *- radical, $k_{ef} = k_r$, if the reaction is limited by a pair of $k_{ef} = k_x$; in the intermediate case k_r and k_x are added by law to the kinetic resistances. Thus, in theory, developed in (Rapoport N.Ya. and all, 1986; Popov A.A. and all., 1987) the rate of intramolecular reactions continue oxidation chains, occurring in isotactic PP is proportional to the concentration of GG *- conformers in the macromolecule. Number of gauche-conformers at the site of a macromolecule depends on the contour length, the distance between the ends of the site. In the PP amorphous phase always there is a distribution on lengths of chains in the intercrystalline regions, hence, on the contour length and concentrations of GG *- conformers. This leads to the presence distribution on values of the constants k_2. Determined from the gross - value of the kinetics of k_2 are effective for the average values distribution. The smallest value of k_2 in the crystalline phase of PP, which have macrochain conformation of the helix. In (Roginsky V.A., 1977; 1982) studied the reaction of the chain oxidation crystal PP-phase at 200-300°C. Estimate of the effective values gave the value of $k_2 = 2.10^{-5}$ s^{-1}, it is 3 orders of magnitude lower than in the amorphous phase k_{2ef} PP. Hence, PP different grades with different molecular weight distribution have different width of the distribution on values of k_2 and values constants k_{2eff}.

2.5 The reactions of accumulation and disintegration of hydroperoxide PP

As has been shown, PP oxidation occurs predominantly intramolecularly, the kinetic chain moves along the macromolecule. Macroradical RO_2^*, formed by the oxidation of polypropylene, reacts with a hydrogen atom from the tertiary C atom located in the β-position relative to the peroxide radical of their molecules. As a result, intramolecular transfer of a macromolecule oxidized PP formed "blocks" of several adjacent OH-groups.

$$\sim CH_2\text{-}C(CH_3)\text{-}CH_2\text{-}C(CH_3)\text{-}CH_2\text{-}C(CH_3)\text{-}CH_2\sim$$
$$\quad | \qquad\qquad | \qquad\qquad | \qquad$$
$$\quad OOH \qquad OOH \qquad OOH$$

However, the low-temperature oxidation of solid polypropylene (70-110°C) proceeds with alternating intramolecular and intermolecular chain transfer. Intramolecular kinetic of extension chains is limited to small parts of the macromolecule with a favorable set of conformations. As a result, blocks of hydroperoxide can be short. In the solid polypropylene has found about 60% of paired units and about 20% of triads, the share of units with a higher number of hydroperoxide groups is small. It should be noted that in other carbon-chain polymers increases the probability of intramolecular reaction at the high rate of conformational motions. For example, in the polymers with a saturated C-C bond (such as

PE), the kinetic chain are transmitted preferably via a carbon atom (in γ-position), in other polymers (with unsaturated bonds, heteroatoms, phenyl rings, etc.) the transfer of kinetic chains are statistically.In oxidizing polypropylene along with the main hydroperoxide:

$$\sim CH_2\text{-}C(CH_3)\text{-}CH_2\sim$$
$$|$$
$$OOH$$

hydroperoxides can be formed type:

$$
\begin{array}{ccc}
 & CH_3 & CH_3 \\
 & | & | \\
\sim CH\text{---}CH\text{---}CH\sim & \sim CH2\text{---}C\text{---}OOH & \sim CH_2\text{---}C\text{---}CH_2OOH \\
|\quad |\quad | & | & | \\
CH_3\ OOH\ CH_3 & H & H
\end{array}
$$

These hydroperoxides are formed as a result of intermolecular transfer of kinetic chains of oxidation. In the PP also form low molecular weight hydroperoxides such as methyl, ethyl, etc. The hydroperoxide, obtained by thermal oxidation of polypropylene, is not an individual compound, but is a combination of $-OOH$ and other oxygen-containing groups, and the concentrations and location of these groups in the polymer are not unambiguous functions of the hydroperoxide concentration.

2.6 The yield of hydroperoxide per mole of absorbed oxygen

Relationship between the concentration of hydroperoxide and absorption rate of oxygen stored in various pressures of oxygen and described by the following empirical equation:

$$W_{O_2}=a[ROOH]^{1/2} + b[ROOH], \tag{48}$$

where $a=k_2(\delta k_d/k_t)^{1/2}$, $b=\sigma k_d[RH]$; k_d – the constant of ROOH decomposition rate; δ - the probability of degenerated branching of kinetic chains;$\delta = \alpha\sigma$, σ-the yield of radicals from cell; α - the yield of hydroperoxide per mole of absorbed oxygen. The rate of polymers oxidation depends on concentration and constant of ROOH destruction. In its turn concentration of ROOH depends on yield per mole of absorbed oxygen. To determine the hydroperoxide yield may be used balance of hydroperoxide in the oxidation of solid polypropylene. For this the obvious fact may be used that at the maximal hydroperoxide concentration the rates of its formation and decomposition are the same.

$$d[ROOH]/dt = \alpha W_{O_2}-k_d[ROOH]^n, \tag{49}$$

where n - the order of decomposition reaction of hydroperoxide. Substituting the experimental values for apparent rate constant of hydroperoxide decomposition k_p, and its maximal concentration at which d [ROOH]/dt becomes zero. Assuming that the initial stage of polyolefin hydroperoxide decomposition obeys the first-order law, we get:

$$\alpha = k_d[ROOH]_{max} / (W_{O_2})_{max} \tag{50}$$

In liquid-phase oxidation of hydrocarbons the rate of oxygen uptake equals the rate of accumulation of hydroperoxide. The yield ROOH per mole of absorbed oxygen $\alpha=1$. In the autoxidation of solid polymer α much less than unity. For isotactic polypropylene, polyethylene, poly-4-methylpentene from (Emanuel N.M., Buchachenko A.L., 1982) a ≈ 0,2-0,5. This means that only 20-50% of the absorbed oxygen passes into the hydroperoxide, the remaining 80-50% goes into the reaction products, bypassing the stage education hydroperoxide. According to (Shlyapnikov Yu.A. and al, 1986) for PP α = 0,05-0,3. There are several reasons for the drop out hydroperoxide per mole of absorbed oxygen. As a result of intramolecular chain transfer to ROOH appears next to the free valence and forms β-hydroperoxyalkyl radical.The interaction of the free valence with a freshly formed hydroperoxide group can occur in three ways, according to reactions (1'), (2') and (3').

$$
\begin{array}{c}
\qquad\qquad\qquad\qquad\;\; \overset{\text{O*}}{|}\quad \overset{\text{OH}}{|}\\
k_d \qquad\quad \rightarrow \sim\text{-CH}_2\text{-C -CH}_2\text{-C - }\sim\;\;(1')\\
\qquad\qquad\qquad\qquad\;\; \underset{\text{OOH}}{\underset{|}{\overset{|}{\text{CH}_3}}}\quad \underset{\text{H}}{\underset{|}{\overset{|}{\text{CH}_3}}}
\end{array}
$$

~-CH₂-C-CH₂ –C*- ~ | k_m → ~-CH₂-C -CH₂-C –CH₂ –C*- ~ (2')

with substituent groups OOH, CH₃, CH₃ ... CH₃, CH₃, CH₃ and OOH OO*

k_k +O₂ → ~-CH₂-C-CH₂-CH - ~ (3') with CH₃ CH₃

The first way to becoming peroksialkil radical would lead to the death of hydroperoxide group (1'), the second and third - the removal of the reaction center and stabilize the group. The yield of the stabilized hydroperoxide is equal to the sum of relative rates of reactions (2') and (3') to the sum of the rates of all three reactions, i.e. the expression for the yield of hydroperoxide (α) in PP is:

$$\alpha = k_k[O_2] + k_m$$

$$k_k[O_2] + k_m + k_d \text{ or}$$

$$1/\alpha = 1 + \{k_d/k_k ([O_2] + k_m/k_k)\} \qquad (51)$$

In the works (Kiryushkin S.G., Shlyapnikov Y.A., 1975; 1986), reaction (1') is considered as " induced" decay of hydroperoxide. The kinetic chain is moved along macromolecules formed alkyl macroradical who takes a step "backwards", reacts with the preceding neighboring hydroperoxide group to form alcohol and alkoksiradikal.

$$
\begin{array}{c}
\overset{\text{CH}_3}{|}\quad \overset{\text{CH}_3}{|}\quad \overset{\text{CH}_3}{|}\qquad\qquad \overset{\text{CH}_3}{|}\quad \overset{\text{CH}_3}{|}\quad \overset{\text{CH}_3}{|}\\
\sim\text{C—CH}_2\text{—C—CH}_2\text{ -C* }\sim \longrightarrow \sim\text{C—CH}_2\text{—C -CH}_2\text{- C }\sim\\
\underset{\text{OOH}}{|}\quad \underset{\text{OOH}}{|}\qquad\qquad\qquad \underset{\text{OOH}}{|}\quad \underset{\text{O*}}{|}\quad \underset{\text{OH}}{|}
\end{array}
$$

It should be noted that "return motion" of the kinetic chain R* + ROOH $\longrightarrow$RO* + ROH , which leads to the induced decomposition of hydroperoxide not necessarily occurs intramolecularly, it can also occur when the intermolecular kinetic extension chains. Only the probability of the latter case will be considerably less than intramolecular reaction, when the local concentration of ROOH and R * is high and partners reactions can not break up the diffusion way. Intramolecular continuation of the kinetic chain of oxidation is not the sole reason for the low yield of hydroperoxide per mole of absorbed oxygen. If the kinetic chain oxidation of long, almost all oxygen passes into the hydroperoxide, with short chains, much of it remains products in the termination of kinetic chains, so α depends on the length kinetic chain. The oxidation of polypropylene in the induction period (long kinetic chain) α = 0.85, almost all the oxygen passes into the hydroperoxide. For short chains of about 30% absorbed oxygen is consumed in the initiation of intracellular processes and termination of kinetic chains, with what shorter kinetic chain, the lower yield of hydroperoxide. On the same reason, the value of α particularly low in crystalline polymers, where kinetic chains are short. Thus low the value of α is not a strict criterion of intramolecular oxidation. It may also be a sign of short kinetic chains. Kinetic chain length (ν) - the number of molecules absorbed oxygen or hydroperoxide at a kinetic chain is

$$\nu = W_{ROOH}/W_i = k_2[RH]k_t^{-1/2} W_i^{-1/2} \tag{52}$$

(Emanuel N.M., Buchachenko A.L., 1982). Another reason low yield of hydroperoxide is the decay of the peroxide macroradical before it turns into a hydroperoxide. In this case, kinetic chain extension occurs without the formation of hydroperoxide. However, the decay of peroxide radicals PP is small compared with the probability of becoming it hydroperoxide. The collapse of the radical RO_2^* can occur at temperatures above 200-300ºC. According to eq.(49) the hydroperoxide yield in the polymer oxidation must depend on the oxygen pressure over the polymer (on its concentration in the polymer). At the same time, oxygen concentration, affecting the ratio of the reaction rates, must affect the hydroperoxide structure, i.e. the distribution of hydroperoxide and other oxygen – containing groups along macromolecules and in polymer bulk. This distribution must affect the hydroperoxide properties. Experiments have proved this conclusion: the rate constant of hydroperoxide decomposition varies depending on the oxygen pressure at which the hydroperoxide has been prepared.

2.7 The probability of degenerate branching

The effectiveness of the emergence of new kinetic chains in the decay of the hydroperoxide has two features. First, the radical yield on one broken hydroperoxide group is small and amounts to several percent. Second, the probability of extinction branching depends on the oxygen pressure. The yield of low molecular weight hydrocarbons (methane, propylene) - the decay products of alkoxide and terminal alkyl macroradicals with an increase in oxygen pressure is not reduced to zero, and reaches a limiting finite value, independent of pressure oxygen. This means that oxygen is not only reduces the concentration of alkyl radicals, converting them into peroxide, but also participates in the reactions of their formation, by increasing output radicals from the cage. Calculated from data on collapse GP isotactic polypropylene dependence of d on the oxygen pressure in 100ºCmeet view d = $(7.10^{-3} +1.10^{-4})$• po_2, where po_2 in mm Hg. Probability degenerate branching of the oxygen pressure for isotactic PP and PE are found from the kinetics of autoxidation is presented in the table.2.

$po_{2,mmHg}$	0	50	100	200	400
σpp	0.017	0.12	0.25	0.40	0.65
σpe	-	-	0.26	-	0.32

Table 2. Probability degenerate branching for isotactic PP and PE

To explain the dependence of δ on oxygen pressure offered two hypotheses. Firstly, the probability of degenerate branching, δ is the initiating ability of ROOH depends on its structure and hence on the oxygen pressure at which the obtained hydroperoxide. Initiation efficiency block hydroperoxides, an order of magnitude higher than the efficiency of initiation isolated groups. However, this hypothesis is realized at low pressures oxygen (less than 150-200 mm Hg). Second, it increases the oxygen probability of degenerate branching root in the primary radical pair converts it into the secondary:

$$
\begin{array}{ccc}
& CH_3 & CH_3 \\
& | & | \\
& \sim C \sim & \sim C \sim \\
& | & | \\
& O^* & O^* \\
(ROOH + RH) \longrightarrow & \quad +O_2 & \\
& \sim C^* \sim & O\!\!-\!\!O^* \\
& | & | \\
& CH_3 & \sim C \sim \\
& & | \\
& & CH_3 \\
& \sigma_o \approx 10^{-3} \div 10^{-2} & \sigma >> \sigma_o
\end{array}
$$

The radical yield of primary pair (σ_o) very low, and from the secondary - much higher, order of unity. The higher the oxygen pressure, the greater the concentration in the polymer, the greater the contribution secondary pairs, and, consequently, the higher the probability of degenerate branching.

2.8 Autoxidation induction period of the PP

The changes, occurring at oxidation of the polymer are laid in the induction period, so it is important establish the nature of this stage of the oxidation process. The parabolic law of oxidation of isotactic PP is established not immediately but after some time. This time was proposed to call the true induction period of uninhibited oxidation (Rapoport N.Ya and all 1986; Popov A.A. and al., 1987). For example, in PP at 403K the true induction period is equal ~ 80% of the experimental period, which correspond to time of absorbed oxygen [No_2] ~ 0,1 mol/kg. Probably mechanisms of oxidation of the polymer in the period induction and on the stage of more extensive oxidation are different. In the literature discusses the possible reasons for the differences of the kinetics of oxidation PP in the induction period, and after leaving it. The following explanation of the features mechanism of oxidation in the induction period are offered in the literature. First, the initial stage oxidation corresponds to the accumulation of hydroperoxide. However, the quasi-steady on hydroperoxide do not run in the induction period. The low yield of hydroperoxide per mole of absorbed oxygen, and low initiating ability of hydroperoxide, which formed at the initial stage, leads to slow down the rate of process of branching of kinetic chains (Popov A.A. and al, 1987). Secondly, the induction period may be associate with the low rate constant of radical decay. The rate constant of radical decay is changed during oxidation. Low - molecular radicals formed at the beginning of oxidation is exchanged for macromolecular radicals in reactions with hydroperoxide (Roginsky V.A. and al., 1976; 1982; Emanuel N.M. and al, 1982). Third, the existence of the period induction may be explained of the localization of the oxidation in the zones (Shlyapnikov. Yu.A. and al 1986; Richters P., 1970; Graeme A.and all., 1997; Livanova, Zaikov G.E., 1997). Localization of oxidation in the zones of polymer is the consequence of the structural and physical microinhomogeneity, nonequivalence of the structural elements , that differ frequencies and amplitudes of molecular motions. This creates spatial heterogeneity in distribution of the reactants in a polymer. Their local concentration may differ significantly from the average. This leads to distribution of reactivity, rate constants and energies activation, as a result, to polychromatic kinetics. This creates spatial heterogeneity in distribution of the reactants in a polymer. This means that oxidative processes is localized to the centers, "Microreactors", which are amorphous interlayers and interfibrillar areas. Polyethylene and polypropylene are not homogeneous. They have amorphous and crystalline regions. In the PE crystalline regions are impermeable and inaccessible for oxygen. The solubility of oxygen in the crystalline regions of PP on order of magnitude smaller than in the amorphous regions. The oxidation rate, calculated on the unit volume, and the limiting amount of absorbed oxygen are decreased proportionally with increasing crystallinity in both polymers (Bogayevskya TA, and al., 1978). Ther are several models of local oxidation of polymers proposed in the literature. In the framework of the local oxidation, polypropylene is considered as a set of kinetically nonequivalent "zones". This zones are differ of molecular dynamics, values of the radical yield of cells, which initiated the kinetic chains, and of termination rate constants k_t. (Makedonov Yu. V. and al, 1986; George A.G.and al., 1997). High molecular mobility, the velocities of the initiation and radical decay in zones leading to rapid establishment of steady-state concentration of radicals in these zones. In the more hard regions, where the rate constants k_t are low and the rate of initiation are small due to cellular effects the process of establishing steady-state concentration of radicals slows down. Therefore, initiating of the radical process in PP by irradiation with light leads to high rates of population of the soft zones of polymer by radicals with high rates of destruction. At longer initiating of the radicals they inhabit the rigid zones, resulting in the experimentally observed rate constant destruction of free radicals decreases. Thus, induction period of oxidation of PP is explained with in terms of non-homogeneous oxidation. By the end of the true induction period is set steady-state distribution of radicals in the zones, the parameters of oxidation are characterize the process of development and not change during further oxidation. Next "zone`s model" of oxidation of the polymer has been proposed (Shlyapnikov and al,1986; 1989). If in the previous model are considered as zones of amorphous regions as a whole, in the second model - it's part amorphous layers in violation of the short-range order. So, the amorphous interlayers of PP include through-passage chains in the folded conformation. In these areas, are concentrated oxygen and other low molecular weight substances. These areas are possess a high segmental mobility and high activity with respect to oxygen. These zones are considered as microreactors, which are surrounded by a more orderly and, therefore, less reactive substance. The model, which offers Shlyapnikov, has different mechanism of oxidation. From the perspective of this model, free valence, formed in the zone of violation of the order, begins a chain reaction, which has no stage of chain termination.The average concentration of RO_2^* in the zone will be equal to one particle per volume area (particles / cm^3 or cm^{-3}.) $[RO_2^*]_z = 1/V_z$. Accepted that V_z = const and we can neglect differences in the individual properties of zones. The expenditure of reactionary capable RH - groups in the some areas (RH_z) are proceedung of the law:

$$d[RH]_z/dt= k_2[RH]_z/V_z.$$
(53)

It concentration of groups in the area of RH, which does not coincide with average concentration of monomeric in the polymer. During $t_z = 5k_2^{-1}V$ reactive substance in violation of the order of the area consumed almost completely. Then there are two possibilities. First, if the average time of free valence in a separate zone t_z much less than θ, then the kinetics of reactions in polymer is not significantly different from the kinetics of the same reaction in low molecular weight liquid only in the reaction will not participate the entire polymer and part of it is equal to $[RH]_e = [Z] V_z [RH]_z$, where Z-density zones. Second, if the stay of free valence in the area of more than θ, then reaction rate is determined by the amount of matter in a separate area and speed displacement of the free valence from zone to zone:

$$W = V_z [RH]_z \theta^{-1} [RO_2^*]$$
(54)

where $[RO_2^*]$ - average concentration of peroxide radicals, calculated on the entire polymer. At the same time the observed rate constant of chain transfer is

$$k_{2e} = V_z [RH]_z [RH]^{-1}\theta^{-1}$$
(55)

and, consequently, will not coincide with the true constant reaction rate RO_2^* + RH, where this constant attribute. The rate of chain termination is determined by the frequency hit two free valences in the same area, i.e. will be:

$$W=2 [Z]^{-1}\theta^{-1} (1-\varepsilon) [RO_2^*]^2,$$
(56)

where ε – chance what a pair of free valences, which has appeared in the same area, come out of it without recombination. Assuming, ε we can write $k_{t\,ef} = 2 [Z]^{-1}\theta^{-1}$ kinetic parameter of oxidation k_{2ef2}/k_{tef} can be:

$$k^2_{2ef}/k_{t\,ef} = V_z^2 [RH]_z^2 N_{av}10^{-6} / 2 [RH]^2\theta$$
(57)

$$\text{where } \theta = V_z [RH]_z [RO_2^*] N_{av} 10^{-3}/W_{o2}$$
(58)

Substituting (56) into (55) we find:

$$k^2_{2ef}/k_{t\,ef} = V_z [RH]_z W_{o2} N_{av}10^{-6} Z / 2[RH]^2 [RO_2^*]$$
(59)

substitution (57) into (35) gives:

$$N_{o2} = \delta k_d [RH] V_z [RH]_z W_{o2} N_{av}10^{-6} Z t^2/16 [RO_2^*]$$
(60)

Under provision, "the zone model" $[RO_2^*] = 1/V_z$, then one can assume that $V_z [RO_2^*]$ = const. Whence

$$N_{o2} = B Z W_{o2} t^2$$
(61)

$$\text{where } B = \delta k_d [RH] V_z N_{av}10^{-3}/16 [RO_2^*]$$
(62)

On the initial stage of oxidation $W_{o2} = N_{o2}/t_{ind}$. Substituting this expression (60) we obtain

$$W_{o2} = B \, N_{o2} \, Z \, t^2/t_{ind} \qquad (63)$$

Simplifying (55), we obtain the desired dependence:

$$t = (t_{ind}/Z)^{1/2} (1/B)^{1/2} \qquad (64)$$

Within the "zone`s model" of the induction period is correlated with the autoxidation of PP time location of the free valence in the area, the low yield of hydroperoxide a can be explained by high local concentrations UN Teams and the radicals R* in the zone, which is why the probability of their interaction is significant. Low yield of free radicals in decay. In within the "zone`s model" of the induction period is correlated with the autoxidation of PP time finding free valence in the area, low yield hydroperoxide can be explained by a high local concentrations of the ROOH and the radicals R * in the zone, which is why the probability their interaction is significant. Low yield of free radicals the decay of the polymer hydroperoxides due to the fact that, unlike low molecular weight Fluid pair of free radicals, emerged from the primary cell is a long time in a small volume of zone violations of the order, where the probability of recombination radicals is high. The low molecular weight scavengers of free radicals dissolved in the polymer are in the same zones violations of the order in which reaction occurs, and may react with the radicals issued from the primary cells, but do not go out of the volume of the zone, are able to initiate the reaction of oxidation. Polymer hydroperoxides due to the fact that in contrast to low molecular weight Fluid pair of free radicals, emerged from the primary cell is a long time in a small volume of zone violations of the order, where the probability of recombination radicals is high. The low molecular weight scavengers of free radicals dissolved in the polymer. Zone`s model allows us to understand the dependence of the oxidation rate related polymers on the content of foreign links. Expression for the effective rate constants of the chain contains as one of the factors value of V_z - volume of the zone violations of short-range order rate constant of quadratic chain termination depends only onthe total concentration of these zones. Introduction to the polymer chain side substituents leads to loosening of the polymer structure and hence, increases the volume of the zone V_z and the rate of oxidation of the polymer. For the same reason, reduces the probability of radical recombination in the area and increases. The transition from PE to PP leads to an increase in the rate of oxidation and reduction induction period (Shlyapnikov Yu. A., 1989). The rate of primary initiation, therefore induction period of oxidation of PP determined not only by the amount and concentration of reactive zones, but also nature of the substances, which are localized in these zones. In the papers convincingly (Livanova, Zaikov G.E., 1997) shown that the polymer (PP) of preexisting primary foci of initiation rate of radicals in which is significantly higher than those in microreactors, which can occur under the influence of stress and other influences on the RH bond PP. The main reason initiation of the primary foci are microscopic debris size (residual polymerization catalysts - transition metal valence and their products conversion).

2.9 Thermooxidative degradation of polypropylene

Hydroperoxide is not only branching agent, but also a precursor of low-molecular products and breaks the molecular chains, resulting in to a change in molecular weight and molecular weight distribution (Shlyapnikov Yu. A., and al. 1986; 1989). Basic mechanisms of decomposition of hydroperoxide in the polypropylene - the bimolecular reaction involving communication C-H at the tertiary carbon atom of the macromolecule eq (8). Among decay products detected ROOH PP: water, acetone, acetaldehyde, formaldehyde, methane, ethane, propane, propylene, ethylene, etc. The rate of formation of these products proportional to the concentration hydroperoxide. Comparative analysis of oxidation products and degradation products hydroperoxide. in an inert atmosphere, showed the same qualitative and quantitative composition. So products oxidation of PP is mainly formed by the decay of hydroperoxides. The total rate of formation of volatile products (of which the main product water-based) half the rate of decomposition of hydroperoxide. Ie 50% of productsoxidation remains in the solid phase in the form of alcohol, peroxide and ketone groups of macromolecules. The main source of low-molecular products are alkoxide macroradicals:

$$\begin{array}{ccc}
CH_3 & & |\rightarrow \sim C^*H_2 + \sim CH_2CCH_3 \\
| & & | \quad\quad\quad \| \\
\sim CH_2-C-CH_2\sim & | & \quad\quad\quad O \\
| & | & \\
O^* & |\rightarrow \sim CH_2CCH_2\sim + (C^*H_3 \rightarrow CH_4) \\
& \| & \\
& O &
\end{array}$$

Low molecular weight hydrocarbons obtained from the terminal alkyl macroradicals:

$$\sim CH_2-CH-C^*H_2 \rightarrow \sim C^*H_2 + CH_3CH=CH_2$$
$$|$$
$$CH_3$$

Isomerization $\sim C^*H_2 \longrightarrow \sim C^* -CH_2CH-CH_3 \rightarrow \sim C=CH_2 + (CH_3)_2C^*H$
$$\quad\quad | \quad\quad | \quad\quad\quad | \quad\quad\quad \downarrow$$
$$\quad\quad CH_3 \quad CH_3 \quad\quad CH_3 \quad\quad CH_3CH_2CH_3$$

Acetone and aldehydes are obtained of hydroperoxide groups and alkoxide radicals, respectively, located near the ends of macromolecules:

$$\begin{array}{l}
CH_3 \\
| \\
\sim CH_2-C-CH_2\sim \rightarrow \sim C^*H_2 + (CH_3)_2CO \\
| \\
O^*
\end{array}$$

$$\begin{array}{ccc}
CH_3 & O & O \\
| & \| & \| \\
CH_3C-CH_2-C-CH_2\sim \rightarrow \sim CH_2-CCH_3 + CH_3C-C^*H_2 \\
\| \quad\quad | & & \downarrow \\
O \quad\quad O^* & & O \\
& & \| \\
& & CH_3CCH_3
\end{array}$$

$$\begin{array}{l}
\sim CH_2-C-CH_2\sim \rightarrow \sim C^*H_2 + CH_3CHO \\
| \\
O^*
\end{array}$$

$$\sim CH_2OOH \longrightarrow \sim CH_2O^* \longrightarrow C^*HCH_3 + CH_2O$$

In polyethylene formation of the products described in the following scheme (Ranby B., Rabek J.F., 1978)

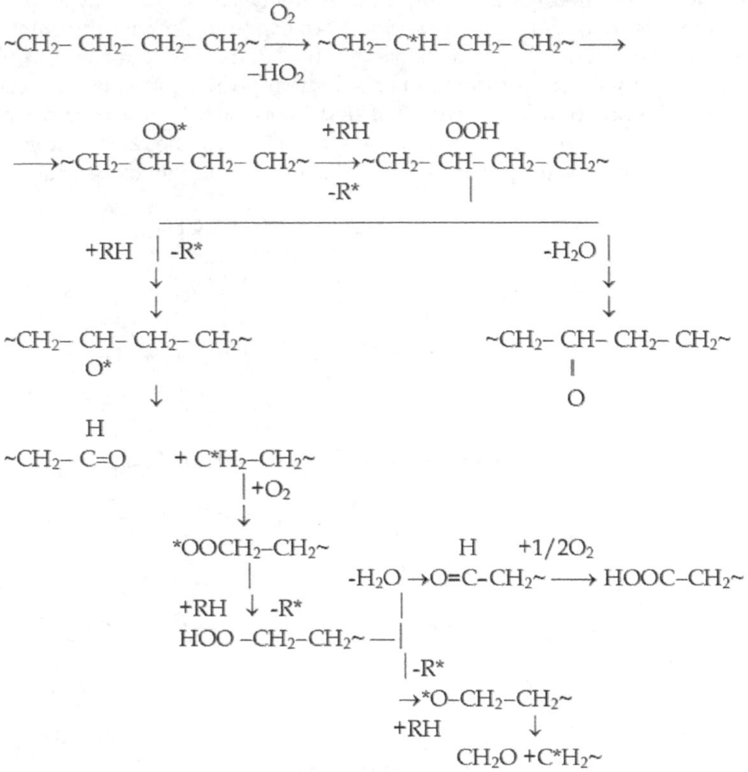

$$\sim CH_2-CH_2-CH_2-CH_2\sim \xrightarrow[-HO_2]{O_2} \sim CH_2-C^*H-CH_2-CH_2\sim \longrightarrow$$

$$\longrightarrow \sim CH_2-\underset{\overset{|}{OO^*}}{CH}-CH_2-CH_2\sim \xrightarrow[-R^*]{+RH} \sim CH_2-\underset{\overset{|}{OOH}}{CH}-CH_2-CH_2\sim$$

$$\begin{array}{cc} +RH \mid -R^* & -H_2O \mid \\ \downarrow & \downarrow \\ \downarrow & \downarrow \\ \sim CH_2-\underset{\overset{|}{O^*}}{CH}-CH_2-CH_2\sim & \sim CH_2-\underset{\overset{|}{O}}{CH}-CH_2-CH_2\sim \\ \downarrow & \end{array}$$

$$\sim CH_2-\underset{\overset{|}{H}}{C}=O \quad + C^*H_2-CH_2\sim$$
$$\mid +O_2$$
$$\downarrow$$

$$\underset{\overset{|}{+RH \downarrow -R^*}}{^*OOCH_2-CH_2\sim} \qquad \underset{-H_2O \to O=C-CH_2\sim}{\overset{H}{}} \xrightarrow{+1/2O_2} HOOC-CH_2\sim$$

$$HOO-CH_2-CH_2\sim -\mid$$
$$\mid -R^*$$
$$\to {}^*O-CH_2-CH_2\sim$$
$$+RH \qquad \downarrow$$
$$CH_2O + C^*H_2\sim$$

Polymer oxidation is accompanied by polymer chain destruction. When measuring the molecular weight of polymer, one can see only large remnants of the polymer chain, and if a short segment of the polymer chain takes place a few breaksthey will be treated as one scission .Denoting the probability of the scission of a macromolecule as a result of one step of chain reaction propagation ρ_1, the probability that there is no scission is $(1-\rho_1)$, and the probability of at least one scission resulting from ν_b reaction steps proceeding in the same region (block) of the macromolecule is (Shlyapnikov Yu. A., 1989):

$$\rho_b = 1-(1-\rho_1)^{\nu_b} \tag{65}$$

The rate of formation of such oxidatized regions (or blocks) of polymer chains is the sum of rates of chain initiation and of chain transition from one molecule to another. Supposing the latter to be directly proportional to the overall concentration of free radicals in the system, gets this:

$$W_b = W_0 + k_{tr} [RO_2^*], \tag{66}$$

where W_0 is the chain initiation rate. It is assumed that the oxygen concentration is high enough and we can neglect reactions of R^* radicals. The rate of recorded scission formation is proportional to W_b, and if $\rho_b=1$ it is virtually equal to W_b. If the oxidation chains are long enough, or k_{tr} is small, the rate of polymer destruction, i.e. the rate of recorded scission formation, will be equal to W_o, i.e. to the rate of chain initiation. The rate of chain initiation plus the rate of chain branching are always equal to the rate of chain termination; thus the rate of chain scission is proportional to that of chain termination, and at ρ_b =1 is approximately equal to it (Shlyapnikov Yu. A. 1986; 1989).

3. The effects of structure on thermooxidation kinetics of polypropylene

The direct correlation between initial structure of polymer and kinetics of oxygen absorption is consequent from our scheme (Shibryaeva L.S.and al.,2003; 2006; 2010.)

MORPHOLOGY OF POLYPROPYLENE
(Isotactic, atactic, sindiotactic PP)
↓
PERMOLECULAR STRUCTURE
(Crystallinity and crystalline size,
The structure of amorphous phases)

↓	↓	↓
Distribution of reagents	**Molecular mobility**	**Contormational set**
↓	↓	↓
/	|	\
Structurally-	**Diffusion**	**Probability of**
orientational correspondence	**of reagents**	**intramolecular reactions**

The thermal oxidation is complex process including chain oxidation of hydrocarbon radicals, destruction of macro-chains and structure formation (cross-linking, crystallization). Thermal oxidation is accompanied by structural-physical processes leading to structure change (structural reconstruction) under the action of high temperature. The mechanism of these processes will depend on polymer's morphology and in its turn will influence on oxidation kinetics. The effect of polymer crystallites in vacuum and on air: the effect of high temperature may lead to perfection of crystallites structure, rise of temperature and melting heat, at the same time at long high temperature effect the destruction of chains occurs and crystallites and decomposed. There are the data demonstrating the influence of annealing temperature on relaxation parameters in polymer which allow concluding that there is significant change of structure of amorphous regions. At that the amorphous regions to a greater extent determine the particularities of oxidation radical reactions kinetics developing in them. With the aim of revealing of the role of structure (conformational set) of polymer macro-chains we also studied structure reconstructions, accompanying oxidation of oriented samples of PP with various extract degree. Structure parameters of PP: crystallites and amorphous regions, make it polymer heterophase system. These parameters of systems will determine localization of oxidation in zones having high segment mobility. In these Section is proposed a model for heterogeneous thermal oxidation of PP. Morphological irregularity of polymer results from the presence of crystalline and amorphous regions in the same polymer. This type of irregularity affects the regularities and rates of polymer oxidation. Crystallites are characterized by long-range order in the arrangement of macromolecules and of their monomeric units. Oxygen solubility in the crystallites is very low or zero, and the R* radicals present in crystalline zones of the polymer cannot transform into peroxide ones. On the other hand, these radicals can move inside the crystals by subsequent reactions R* + RH. The capture of free radicals R* by crystallites is equivalent to the kinetic chain termination if these radicals remain in the crystallites or recombine in them. It was shown (Shlyapnikov Yu.A., 1986) that if these radicals are only kept inside the crystallites for a certain time, this is equivalent to chain termination if the reaction is self – accelerated.

4. Isotactic pp, modified by oil

The given section is devoted to regularities of thermal oxidation and to structural reconstructions in the course of oxidation of model heterogeneous systems (Shibryaeva L.S., 2010). The change of destruction rate of PP chains in the presence of modifier may be explained by the change of mechanism of polymer oxidation. Increase of segment mobility of chain leads to increase of contribution of intermolecular transfer of kinetic chains of oxidation. As a result of this kinetic chains of oxidation become shorter and the number of breakages in macromolecules is increased. But in the case of destruction process acceleration at the stage of kinetic chains continuation at the expense of prevailing of intermolecular transfer in composition of hydroxyl containing products the single OH-groups will prevail. However analysis of products composition did not reveal prevailing of single OH-groups over block ones. Increase of chains mobility is observed not only in the case of PP with compatible additives, but also for uncompatible, nevertheless the rate of PP oxidation in its presence is reduced. The most probable reason of oxidation process acceleration in samples of PP with compatible additives, of the rise of PP chain destruction rate is joint oxidation (co-oxidation) of polymer and additives. There are two fundamental hypotheses concerning the mechanism of these reactions which can be derived from the present experimental knowledge of structure and reactivity of macroradicals. (1) Hypothesis of physical migration by which the reactions of reactants are controlled either by mobility of the chains or their parts (segments) with fixed free valence or by diffusion of the low-molecular products of macroradical decomposition, e.i. the so-called radical fragments. (2) Hypothesis of chemical migration by which the reactions of reactants are controlled by various reactions of macroradicals of low-molecular fragments.

5. Experimental part

The samples of Isotactic PP non-inhibited powder of polymerizate was purified by standard technique.

5.1 The methods of investigation

5.1.1 Kinetics of oxidation

The kinetics of oxidation of isotactic polypropylene was investigated in circulating apparatus with freeze-volatile products of oxidation at the temperature of liquid nitrogen. When the film thickness is less than 60 microns maximum rate of oxidation of the sample is proportional to its thickness. Consequently, at a thickness of less than 60 microns (l<60 mcm) kinetic regime is realized, i.e. diffusion of oxygen is a rapid process and does not affect the rate of oxidation. On the contrary, for l > 200 microns oxidation occurs in the diffusion regime and the maximum speed calculated for 1 cm^2 the surface is practically independent of the thickness.

5.1.2 The research methods

Differential scanning calorimetry, X-ray analysis, IR-spectroscopy, Electronic microscopy, ESR-study.

6. List of abbreviations

PE	- polyetylene
PP	- polypropylene
PEHD	- polyethylene of high density
PELD	- polyethylene of low density

7. References

Bogayevskya TA, Monakhova TV, & Shlyapnikov YA (1978) "On the connection between the crystallinity of polypropylene and the kinetics of the oxidation" *Vysocomolek. Soedin.* V.20 B. № 6.PP.465-468.

Chan J.H., & Balke S.T. "The thermal degradation kinetics of polypropylene. *Polymer Degradation and Stability* (1997). V. 57.№ 2. Part. I. Molecular weight distribution. PP 113-125. Part II. Time-temperature superposition. PP.127-134. Part III. Thermogravimetric analyses. PP.135-149.

Denisov E.T., Miscavige N.I., & Agabekov V.E..(1975). *The mechanism of liquid-phase oxidation of oxygen-containing compounds.* Minsk: Science and Technology. P. 200.(In Russian)

Emanuel N.M., Buchachenko A.L.. (1982). *Chemical Physics of aging and stabilization of polymers.* Moscow: Nauka. P. 360. (In Russian)

George A.G., Celina M., & Lerf C. at al. (1997) "A Spreading Model for the Oxidation of Polypropylene" *Makromol. Chem., Macromol. Symp.*V 115. № 1. PP.69-92.

Livanova N.M., & Zaikov G.E. (1997) "The Initiation of Polypropylene Oxidation"*Intern. J. Polymeric. Mater.* V.36. №1. PP.23-31.

Makedonov Yu.V., Margolin A.L., & Rapoport N.Ya. and al. (1986)." On the causes of changes in the effective rate constants continue and chain termination in the induction period of oxidation of isotropic and oriented isotactic polypropylene". *Vysocomolek. Soedin.* V.28 A. № 7. PP.1380-1385. (In Russian)

Nakatani H., Suzuki S., & Tanaka T. and al. "New kinetic aspects on the mechanism of thermal oxidative degradation of polypropylenes with various tacticities". (2005). *Polymer.* V.46. № 11.PP.12366-12371.

Popov A.A, Rapoport N.Ya, & Zaikov G.E.. (1987). *Oxidation-oriented and stressed polymers.* Moscow: Khimiya. P. 168. (In Russian)

Ranby B.,& Rabek J.F (1978). *Photodegradation, Photo-oxidation and Photostabilization of Polymers.* Moscow: Mir. (In Russian).

Rapoport N.Ya., Mostovaya E.M., & Zaikov G.E. (1986)."Simulation of Polychromatic Kinetics of the Propagation step of the Chain Oxidation Reaction in Strained Polypropylene Basing on the Rotational Isomers Model " *Vysokomolek. Soedin.* 28 A. № 8. PP.1620 -1628. (In Russian)

Roginsky, VA, Shanina EL,& Miller, VB (1976) "The cellular effect of the decay of the initiators in solid polypropylene" *Dokladii of the Academy of Sciences of USSR* V. 227.

№ 5. PP.1167-1170.; (1982) "The loss of peroxide macroradicals in the crystalline phase of isotactic polypropylene". *Vysokomolek. Soedin.*V. 24. № 6. PP.1241-1246.

Shibryaeva L.S., Shatalova O.V., & Krivandin A.V. and al. (2003) "Structure Effects in the Oxidation of Isotactic Polypropylene" *Vysokomolek. Soedin.* 45 A. № 3.PP.421-435.

Shibryaeva L.S.,Popov A.A., & Zaikov G.E. (2006). *Thermal Oxidation of Polymer Blends.* Ch.2"Structure effects in thermal oxidation of polyolefines".Leiden, Boston: VSP P.35-55. ISBN -13: 978-90-6764-451-8; ISBN-10:90-6764-451-X.

Shibryaeva L.S., Shatalova O.V., & Solovova J.V. and al. (2010). "Features of thermal oxidation of polypropylene modified with paraffin oil." *Theoretical osnovy himicheskoy technologii.* V.44. № 4. P. 454-466.

Shlyapnikov, Yu.A., Kiryushkin S.G & Marin. A.P.1986. *Antioxidantive stabilization of polymers.* Moscow. Chemistry.P. 254. (In Russian).

Shlyapnikov Yu.A. (1989) "Specific features of polymer oxidation" *Makromol. Chem., Macromol. Symp.* V.27. № 1. PP. 121-138.

4

Mechanical Behavior Variation of an Isotactic Polypropylene Copolymer Subjected to Artificial Aging

Hugo Malon, Jesus Martin and Luis Castejon
University of Zaragoza
Spain

1. Introduction

Most of the injuries suffered by motorcyclists and cyclists on Spanish highways are due to collision with guardrails, particularly with the posts that support these protective devices.

In order to lessen the effect of such accidents, the research group "New Technologies applied on Vehicles and Road Safety" (VEHIVIAL) of the University of Zaragoza is carrying out a series of studies, in collaboration with the company Taexpa S. A.

These studies are developing an innovative a system that will protect motorcyclists and cyclists from crashes against guardrails.

The material chosen for the development of this absorption system is commercial isotactic polypropylene copolymer, due to its great capacity to transform the kinetic energy of the shock into strain energy.

The developed system can be installed on the posts of the highways guardrails, where the polypropylene is affected by the environment, modifying the material mechanical properties.

The isotactic polypropylene is found in almost all of the polypropylene market. It is a crystalline thermoplastic material, so mechanical properties mainly depend on their molecular structure, their crystal structure and the macro-structure induced by the transformation process (Monasse & Haudin, 1995; Rodriguez et al, 2004; Varga, 1992; Fujiyama et al, 2002).

The chapter shows the analysis of the mechanical properties variation of an isotactic polypropylene copolymer, when this material is subjected to artificial aging according to the standard UNE 4892.

The selected material is polypropylene because it is the polymer that has a better impact resistance-price ratio.

2. Materials and experimental techniques

In order to obtain the variation of the mechanical behaviour of polypropylene copolymer under study, experimental and virtual testing techniques have been used.

In the first phase of the developed process a series of samples was subjected to artificial aging in a climatic chamber. The mechanical properties of the original material used and the aged polypropylene were obtained by tensile tests.

Once the mechanical properties of both materials were obtained, a series of numerical calculations by means of the Finite Elements Method (FEM) were made. The results of the numerical tests allow for obtaining the variation of the mechanical behaviour of the material subjected to artificial aging with respect to the original polypropylene.

2.1 Materials

The material used in the study is an isotactic polypropylene copolymer, because it is the polymer that has a better impact resistance-price ratio. To carry out the experimental tests 10 dumbbell-shaped samples of type 1B were mechanized from a square sheet of side 1000 mm and a thickness of 3 mm of this material .These samples were manufactured according to the standard UNE-EN-ISO 527-2.

2.2 Artificial aging of isotactic polypropylene copolymer

The aim of this test is to reproduce the effects that occur when materials are exposed to the environment.

Artificial aging tests of materials were carried out in laboratory under more controlled conditions than in the natural processes of aging. These tests were designed to accelerate the degradation of the polymer and the material failures.

For this, 5 samples were subjected to cyclical periods of UV exposure, followed by periods without radiation. During these cycles, changes in temperature and humidity were carried out according to the standard UNE 4892. This standard is the one governing the artificial aging tests and it was used as a guide in the investigation developed.

The artificial aging cycle applied consists in two hours of UV exposure at a temperature of 30°C and a relative humidity of 62%, followed by one hour of condensation without radiation at a temperature of 40°C and a relative humidity of 90%. The duration of artificial aging process developed had been 78 hours, which corresponds to 26 cycles.

The equipment used in carrying out the artificial aging process is a climatic chamber Ineltec CC-150, which is located in the premises of the Department of Science and Technology of Materials and Fluids of the University of Zaragoza. The climatic chamber capacity is 125l and the maximum and minimum working temperatures of the machine are 150°C and 10°C respectively.

Figure 1 shows the inside of the climatic chamber with the samples subjected to artificial aging. Figure 2 shows the position of the samples in the climatic chamber, which were positioned in the irradiation zone of the lamp. This lamp is represented in the figure 2 by an X.

2.3 Tensile test of mechanized samples

Once the samples of isotactic polypropylene copolymer were subjected to the artificial aging process, the next phase in the development process was to obtain and compare the mechanical properties of the aged material and the mechanical properties of the original material.

Fig. 1. Inside the climate chamber with 5 samples

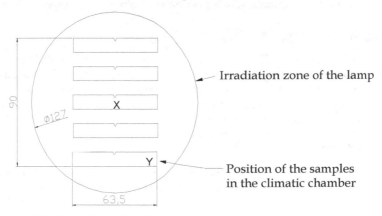

Fig. 2. Position of the samples in the climate chamber

The tensile test is the testing to obtain information of the mechanical properties of materials. The aim of this test is to obtain the elastic modulus, stress-strain curve, yield strength, tensile strength and elongation of a material.

In order to obtain these values, 5 samples with artificial aging (A1-5) and 5 samples without aging process (N1-5) had been tested according to the standard UNE-EN ISO 527 - 1996.

Tensile tests were carried out in an Instron 8032 testing machine at the Department of Mechanical Engineering at the University of Zaragoza. This testing machine has a load capacity of 100kN.

2.4 Numerical analysis by means of the Finite Elements Method (FEM)

The last phase of the study was to obtain the variation of the mechanical behaviour of the material subjected to artificial aging with respect to the original polypropylene in an impact.

The impact analysis in polymers has a number of difficulties (Kalthoff, 1993; Richardson & Wisheart, 1996; Moyre et al, 2000; Read et al, 2001; Tarim et al, 2001; Dean & Wright; 2003; Trudel-Boucher et al, 2003; Jimenez et al, 2004; Aretxabaleta et al, 2004; Aretxabaleta et al, 2004; Aretxabaleta et al, 2005; Davies et al, 2005; Alcock, 2006; Martinez et al, 2008; Aurrekoetxea et al, 2011), which complicate the analysis of such situations. The causes of this complexity have been summarized in the following points:

- The dynamic nature of the problem, including the phenomena of wave propagation Justify, single space (Aita et al, 1992; Bigi,1998)
- The three-dimensionality of the problem, often asymmetric and two-dimensional simplifications being insufficient (Wierzbicki, 1989)
- The behaviour of materials at extreme loads caused by an impacts is almost always non-linear (Krieg & Key, 1976)
- The simulation of material behaviour must be representative of the entire range of strains rates that develop in the impact (Hull, 1985)

The numerical analysis by means of the FEM is a tool to provide solutions to the problems described. This numerical technique has been used and validated in previous studies (Martin et al, 2007). In these previous studies, reliable results were obtained in polypropylene impact.

2.4.1 Application of the FEM

The test to reproduce numerically, by means of the FEM, is the freefall impact of a steel semisphere of diameter 25mm against a square sheet of side 110mm and a thickness of 3mm of polypropylene. This model was validated in previous studies to analyze impact polypropylene sheets. The sheet has been discretized with shell elements of 4 nodes (S4R) and the semi-sphere has been discretized with volumetric elements of 6 or 8 nodes (C3D6 o C3D8R). Figure 3 shows the sheet in green and the semisphere in red. The model consists of 5352 elements and 5568 nodes.

The mechanical properties used in the definition of polypropylene were the averages of the results obtained in tensile tests carried out on the samples subjected to artificial aging and the initial polypropylene samples. The material of the semi-sphere was defined as a linear steel with a Young modulus E = 210GPa, density γ = 7850 kg/m^3 and Poison ratio υ = 0.3.

The methodology applied in the virtual test development has been based on the application of numerical techniques by means of the Finite Element Method (FEM) with the explicit integration of a dynamic equilibrium equation.

Changes in Kinetic energy and strain energy, as well as the displacements in the sheet, are obtained from the simulations for the material with and without artificial aging.

These results provide important information in the optimization process of the developed protection system, because they indicate how the material behaviour changes with the environment, and how it affects the functionality of the energy absorption system developed.

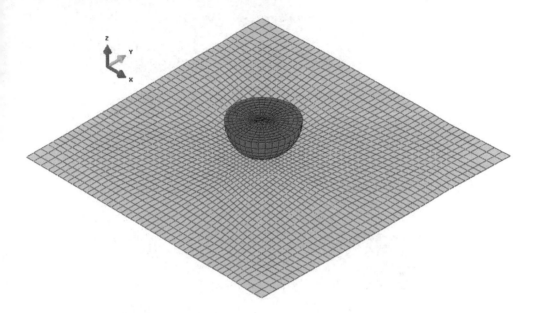

Fig. 3. Finite elements model

2.4.2 Load cases and boundary conditions

The load cases analyzed corresponds to the freefall impact of a steel semi -sphere of 25 mm diameter against a polypropylene sheet of a thickness of 3mm.

In order to obtain greater reliability in the results, three load cases have been analyzed. The difference in load cases is the difference of the height at which the semi-sphere drops.

In order to reduce the computational cost (computation time) of the simulations, the simulation was not carried out on the total trajectory of the semi sphere. Instead, the speed of the semi sphere in the instant previous to the impact was calculated. With this speed and in that position of the semi sphere, the numerical simulation test was initiated. This technique allows for saving the computation time in which the semi-sphere covers the distance from the initial height of the test to the instant previous to the impact with the sheet.

The speed at the instant previous to the impact was obtained by an energy balance, in which at the initial instant of the test, the kinetic energy of the semi sphere is zero ($v_1 = 0$) and at the instant previous to the impact, the potential energy of the semi sphere is zero ($h_2 = 0$). Therefore the potential energy of the sphere at the initial instant of the test is transformed into kinetic energy of the semi sphere in the instant previous to the impact. The following equations show the process developed in order to obtain the speed of the semi sphere at the instant previous to the impact against to the polypropylene sheet.

$$EC_1 + EP_1 = EC_2 + EP_2 \qquad (1)$$

$$\frac{1}{2} * m_1 * v_1^2 + m_1 * g * h_1 = \frac{1}{2} * m_2 * v_2^2 + m_2 * g * h_2 \qquad (2)$$

$$\frac{1}{2} * m_1 * 0 + m_1 * g * h_1 = \frac{1}{2} * m_2 * v_2^2 + m_2 * g * 0 \qquad (3)$$

$$m_1 * g * h_1 = \frac{1}{2} * m_2 * v_2^2 \qquad (4)$$

$$v_2 = \sqrt{2 * g * h_1} \qquad (5)$$

Figure 4 shows a diagram of the starting position and the instant previous to the impact

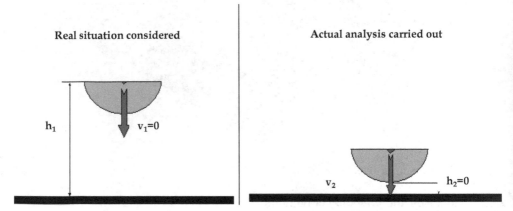

Fig. 4. Diagram of the starting position and the instant previous to the impact

The load cases analyzed are shown in Table 1. The table shows the initial height in the test and the velocity of the semi sphere at the previous instant of the impact.

Load case	Initial height (mm)	Impact velocity (m/s)
1	1,575	5.56
2	790	3.94
3	527	3.21

Table 1. Initial height and impact velocity of the load cases

The imposed boundary conditions reproduce those of a freefall impact test, in which the contour of the sheet is fastened. In virtual testing, rotations and displacements were constrained in nodes located less than 10mm from the edge of the sheet, which are shown in red in Figure 5.

3. Results

Tensile tests carried out on samples of original polypropylene and polypropylene subjected to artificial aging had provided force-displacement curves of the materials, which are shown in figures 6 and 7.

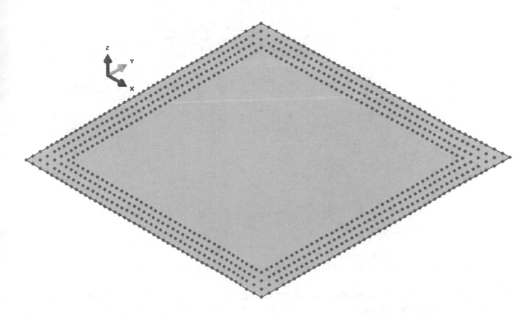

Fig. 5. Nodes with displacements and rotation constrained by the boundary conditions

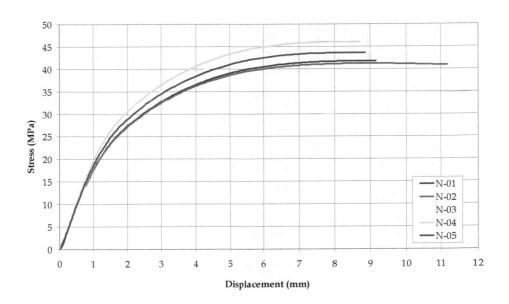

Fig. 6. Stress-displacements curves of the original polypropylene samples

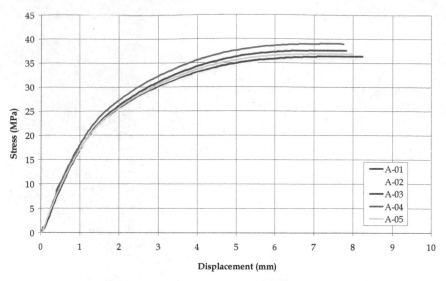

Fig. 7. Stress-displacements curves of the polypropylene subjected to artificial aging samples

The results of tensile tests recorded allow for obtaining the mechanical properties of the materials studied. The Young's modulus of both materials was calculated from σ_1 (stress at a strain of 0.0005) and σ_2 (stress at a strain of 0.0025) according to the standard UNE-EN ISO 527-1. Tables 2 and 3 shown the Young's modulus and the tensile strength obtained of the tensile tests. The other mechanical properties used were density γ = 7850 kg/m³ and Poison ratio υ = 0.3.

Non-aged samples	E_t [MPa]	R_m [MPa]
N1	671,160	41,732
N2	718,130	41,177
N3	841,690	44,369
N4	744,030	46,011
N5	746,610	43,563
Mean	744,324	43,370
Standard deviation	55,72	1,77

Table 2. Mechanical properties of non-aged polypropylene

Aged samples	E_t [MPa]	R_m [MPa]
A1	717,650	38,338
A2	724,760	36,001
A3	652,320	36,472
A4	688,720	37,025
A5	805,950	37,714
Mean	717,880	37,110
Standard deviation	50,89	0,839

Table 3. Mechanical properties of aged polypropylene

Once the virtual tests through the MEF had been run, the analysis of the mechanical behaviour variation of the polypropylene subjected to artificial aging with respect to the original material began. First, the maximum vertical displacements in the sheet were compared for each of the three load cases analyzed. In all cases the maximum vertical displacement and permanent deformation were higher in samples subjected to artificial aging. The difference of vertical displacement and permanent deformation between sheets of material subjected to artificial aging and sheets of original material was greater on increasing initial height of the test. These results are shown in figure 8.

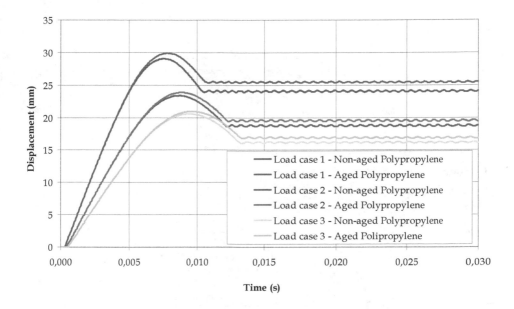

Fig. 8. Maximum vertical displacement on the sheet

The second parameter analyzed is the kinetic energy of the semi sphere, figure 9. The results obtained show that the kinetic energy of the semi sphere after the impact against the sheets of polypropylene subjected to artificial aging is lower than in the impact against the original material sheets for the three load cases analyzed. This greater reduction of the kinetic energy implies a lower speed of the semi sphere in the simulations with polypropylene subjected to artificial aging with respect to the simulation with original material.

The third parameter analyzed is the strain energy, figure 10. This parameter represents the energy used in the deformation of the Polypropylene sheet on the impact. The results obtained show that the energy used in the deformation of the polypropylene sheet subjected to artificial aging is higher than in the original material for the three load cases analyzed.

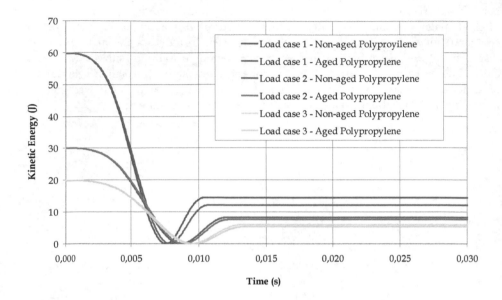

Fig. 9. Kinetic energy of the semi-sphere

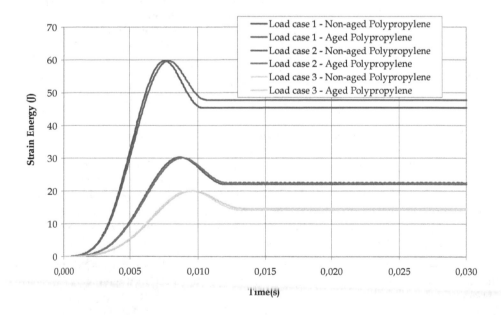

Fig. 10. Strain energy in the sheet

Table 4 summarizes the results of maximum displacements and strain energy in the sheet and the kinetic energy of the semi sphere obtained of the virtual simulations by means of the MEF.

Load case		Non-aged Polypropylene	Aged Polypropylene
1	Initial Kinetic Energy (J)	59,85	59,85
	Final Kinetic Energy(J)	14,44	12,12
	Reduction of Kinetic energy (%)	75,87	79,75
	Reduction of Kinetic energy (J)	45,41	47,73
	Final Strain Energy (J)	45,34	47,72
2	Initial Kinetic Energy (J)	30,05	30,05
	Final Kinetic Energy(J)	8,32	7,63
	Reduction of Kinetic energy (%)	72,31	74,61
	Reduction of Kinetic energy (J)	21,73	22,42
	Final Strain Energy (J)	21,21	22,35
3	Initial Kinetic Energy (J)	19,95	19,95
	Final Kinetic Energy(J)	6,02	5,39
	Reduction of Kinetic energy (%)	69,82	72,98
	Reduction of Kinetic energy (J)	13,93	14,56
	Final Strain Energy (J)	13,86	14,25

Table 4. Results obtained of the virtual simulations by means of the FEM

4. Conclusions

The research process developed allows for obtaining the mechanical properties variation of isotactic polypropylene copolymer subjected to artificial aging.

For this, 10 samples were machined from a sheet of polypropylene copolymer. Five samples were subjected to artificial aging in a climatic chamber.

Subsequently, 10 samples (5 of material subjected to artificial aging and 5 samples of original polypropylene) were subjected to tensile tests in order to obtain the mechanical properties of the original polypropylene and the material subjected to artificial aging.

After obtaining the mechanical properties, numerical analysis by Means of the Finite Element Method (FEM) with explicit integration of dynamic equilibrium equation was carried out. These numerical techniques allow for obtaining reliable results of impacts against polypropylene sheets. Virtual simulations allow for obtaining the maximum displacements in the sheets, the kinetic energy reduction of the semisphere and the energy absorbed by the sheet in the load cases analyzed.

The results show a mechanical behaviour similar to the material subjected to artificial aging with respect to original polypropylene in all the load cases analyzed. Moreover, in all the load cases analyzed the sheets of the material subjected to artificial aging reduce the kinetic energy by a greater amount with respect to the sheets of the original polypropylene. Thus, artificial aging improves the behaviour of the material for use in energy absorption systems.

The minimal variations obtained of the mechanical properties of polypropylene subjected to artificial aging with respect to the original material show that the polypropylene is a suitable material for the design of systems to protect motorists and cyclists. These protection systems are continually exposed to environmental effects, and therefore a continuous aging process.

5. References

Aita, S., El-Khaldi, F., Fontaine ,L., Tamada, T, & Tamura, E., (1992) Numerical Simulation of Stretch Drawn Autobody. Part 1:Assessment of Simulation Methodology and Modelling of Stamping Components,1992 *Proceedings of SAE International Congress 1992*, Detroit,USA

Alcock B. (2006) Low velocity impact performance of recyclable all-polypropylene composites. *Compos. Sci. Technol.*, 66(11–12), pp. 1724–1737.

Aretxabaleta, L., Aurrekoetxea, J., Urrutibeascoa, I. & Sánchez Soto. M. (2004) Caracterización de materiales termoplásticos sometidos a impactos de baja velocidad, *2° Congreso nacional de jóvenes investigadores en polímeros*, Zarauz (Guipúzcoa), 6-10 junio

Aretxabaleta, L., Aurrekoetxea, J., Urrutibeascoa, I. & Sánchez Soto M. (2004) Caracterización a impacto de plásticos: modelos con criterio de fallo, *Anales de mecánica de la fractura*, Vol. 21, pp.310-314

Aretxabaleta L., Aurrekoetxea J., Urrutibeascoa I.& Sánchez-Soto M. (2005) Characterisation of the impact behaviour of polymer thermoplastics. *Polymer Testing*, Vol.24, pp.145–151.

Aurrekoetxea J.; Sarrionandia M.; Mateos M & Aretxabaleta L. (2011) Repeated low energy impact behaviour of self-reinforced polypropylene composites. *Polymer testing*. Vol.30, Issue. 2, pp. 216-221.

Bigi, D., (1988) Simulazione numerica dei problemi di crash veicolistico. *ATA Ingegneria automotoristica*, Vol 41, nh.5, pp386-392.

Davies G.A.O., Zhang X., Zhou G. & Watson S. (2005) Numerical modelling of impact damage. *100th Anniversary Conference and Exhibition of the Centre-for-Composite-*

Materials, Imperial College of Science Technology and Medicine. Composites Vol. 25 Issue.5, pp. 342-350

Dean, G. & Wright, L., (2003) An evaluation of the use of finite element analysis for predicting the deformation of plastics under impact loading, *Polymer Testing,* VOl.22, pp.625-631

Fujiyama, M., Kitajima, Y. & Inata, H. (2002) Structure and properties of injection-molded polypropylenes with different molecular weight distribution and tacticity characteristics. Journal of applied polymer Science. Vol.84 (12), pp.2142-2156

Hull,D., (1985) Impac Response of Structural Composites, *Composite Materals*, pp.35-38

Jiménez, O., Sánchez-Soto, M., Santana, O. O., Maspoch, M. Ll., Gordillo, A.,. Velasco. J. I & Martínez. A. B., (2004), Identación por impacto de baja energía: modelo completo, *Boletín de la Sociedad Española de Cerámica y Vidrio*, Vol.43, pp. 324-326

Kalthoff J.F. (1993) On the validity of impact energies measured with polymeric specimens in instrumented impact tests. *Impact Dynamic Fract Polym Compos ESIS,* Vol. 19.

Krieg, R.D. & Key, S.W. (1976) Implementation of a time dependent plasticity into structural computer programs. Constitutive equations in Viscoplasticity: Computational and Engineering aspects, Vol.20

Martin, J., Malon, H. & Castejón, L. (2008) Validation of the Finite Element Method Applied to Isotactic Polypropylene Homopolymers. *Polymers & Polymer Composites,* Vol.16(0), pp.457-463.

Martínez A.B., Velasco J.I., Gordillo A., & Jiménez, O. (2008) Impacto de baja energía en polímeros y composites. *IX Simposio Latinoamericano de Polímeros/VII Congreso Iberoamericano de polímeros.* Valencia (Spain).

Monasse, B., & Haudin, J.M. (1995). *Popypropylene: Structure, Blends and Composites, Vol I, Structure and Morfology,* Ed.Chapman .Hall, London, pp.3-30.

Morye S.S., Hine P.J., Duckett R.A., Carr D.J. &. Ward I.M (2000), Modelling of the energy absorption of polymer composites upon ballistic impact. *Compos Sci Technol,* Vol. 60, pp.2631-2642

Read B.E., Dean G.D, & Wright L. (2001) Modelling non-linear stress-strain behaviour of rubber-toughened plastics. *Plastics, Rubber and Composites,* Vol. 30, pp. 328.

Richardson M.O.W. & Wisheart M.J. (1996) Review of low-velocity impact properties of composite materials. *Composites: Part A,* 27A, pp. 1123–1131

Rodríguez, S., Perea, J. M., & Vargas, L. (2004) Modelización de propiedades mecánicas del polipropileno: Parte I, grados de síntesis. *Revista de plásticos modernos: Ciencia y Tecnología de polímeros*, n°. 573, pp. 257-262, *ISSN 0034-8708*.

Tarim N., Findik F. & Uzun H. (2001) Ballistic impact performance of composite structures. *Compos Struct,* Vol. 56, pp. 13–20.

Trudel-Boucher D., Bureau M.N., Denault J, & Fisa B. (2003) Low-velocity impacts in continuous glass fiber/polypropylene composites. *Polymer Composites* 24(4), pp.499–511.

Varga, J.(1992) Review: Supermolecular structure of isotactic polipropylene, *Journal of Materials Science* Vol.27, pp. 2557-2579

Wierzbicki, T., (1989) *Geometrical Modeling for Crash*, Post Symposium Short Course of the 2nd International Symposium of Plasticity, Nogoya Agosto1989

5

Rheological Behaviour of Polypropylene Through Extrusion and Capillary Rheometry

Zulkifli Mohamad Ariff[1], Azlan Ariffin[1],
Suzi Salwah Jikan[2] and Nor Azura Abdul Rahim[3]
[1]Universiti Sains Malaysia,
[2]Universiti Tun Hussein Onn Malaysia,
[3]Universiti Malaysia Perlis,
Malaysia

1. Introduction

For the past decade, polypropylene (PP) has become one of the most widely use polyolefin especially for intensive activities in research, product development and commercialization. The factor is strongly contributed by the high demands and usage predominantly in food packaging, automotive industries, fabrication of electric and electronic components and currently its utilization in building structural component for civil needs. Despite their variety in terms of its applications, PP has to be brought into melted stage first before it can be transformed into desired shapes. PP possesses relatively high melting point, low density, high tensile modulus and it is relatively low-priced compared to other thermoplastics. In addition, commercial PP consists of generally linear molecular structure thus it can provide low melt strength and exhibits no strain hardening behaviour in the melted stage. This suggests that PP is suitable for injection moulding, blow moulding and extrusion processing techniques (Rahim[a] et al., 2011).

During the melt processing, polymeric materials are subjected to various rigorous conditions such as high shear deformation and relatively high temperature that could trigger chemical transformation which subsequently leads to molecular degradation and structural development as well as flows through narrow and complex geometries (Ariff, 2003). These factors will definitely influence the rheological behaviour or to be specific the melt viscosity of PP. If the viscosity of molten PP is not suitable within the processing conditions, short shot or flashing may occur in injection moulding. These problems are essentially crucial to be addressed since the modification on thermoplastic viscosity will also modify the end properties of the produced product. Upon the conversion, there are discrete or usually combinations of chemical and physical changes taking place such as chemical reaction, flow or a permanent change which will directly affect the product end properties. With regards to its flow behaviour, a better understanding of PP rheological characteristic can overcome the existing difficulties and literally secure a successful processing.

Moreover, because of the special characteristic (i.e. viscoelastic behaviour) that PP owns, the processing operations are usually more complex than mechanical of chemical engineering unit operations. In PP melt, both viscous and elastic component will deform and this would bring about scenarios that are associated with elastic respond of the polymeric materials which are commonly known as melt fracture or flow instabilities. These phenomena can be easily demonstrated by extruding PP melts through capillary or extrusion die above and below the critical wall shear stress. Another common elastic effect occurring in polymer processing besides flow instability is extrudate swell. It is the most common defect found during extrusion process where the diameter of the extrudate appears to be larger than that of the die (Ariffin et al., 2006). Nevertheless, the knowledge of PP rheological behaviour will equip manufacturers with better quantitative processing responses to cater the actual processing complexity.

2. The importance of polymer rheology

Before PP is turned into product, it will undergo fabrication processes that involve deformation and flow which can be said to be the essence of rheology. Polymer rheological data is used in determining whether or not a type of polymer can be extruded, moulded or shaped into a practical and useable product. Having knowledge of polymer rheology would probably help in determining the optimal design of processing equipment such as extrusion die design, screw geometries of an extruder, various mould cavities for injection moulding and mixing devices. This indicates that in polymer processing operations, an understanding of polymer rheology is the key to efficient design, material and process selection, efficient fabrication and satisfactory service performance. Fig 1 further shows numbers of area where a better understanding of polymer rheology can lead to successful polymer processing operation. (Han, 1976)

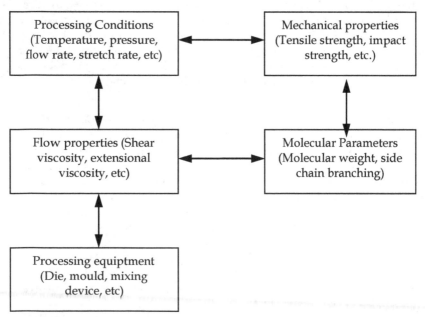

Fig. 1. Schematic of the interrelationship that exists between the processing variables and flow properties, and between the molecular parameters and flow properties (Han, 1976)

Currently there are numbers of methods and instruments available to measure the rheological data of polymers. Since PP is a pseudoplastic thermoplastic fluid that requires relatively high processing temperatures to achieve suitable melt viscosity, it demands a rheological measurement instrument that is able to operate at such temperature under wide of shear rate range. In view of this, capillary rheometer is the simplest and most popular rheological instrument used to measure PP rheological properties (Rahim, 2011; Ariffin et al., 2008; Liang and Ness, 1998; Muksing et al., 2008). This is due to the fact that it has number of advantages over other types of rheometers. First, it is relatively easy to fill which is crucial when dealing with high viscous melt at high processing temperature. Second, the shear rate and flow geometry are similar to the actual condition found in processes such as extrusion and injection moulding (Gupta, 2000). Basically, there are two types of capillary rheometers to measure the viscosity of molten PP namely the pressure-driven type (constant shear stress) and the piston driven type (constant shear rate), yet the approach of their measurement procedure is quite similar where it depends on the applied force (Fig. 2).

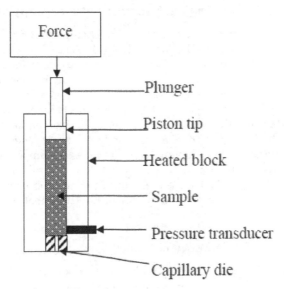

Fig. 2. Schematic diagram for capillary rheometer

The deformation and flow from shear mechanism introduced during polymer processing will results in molecular orientation which creates a dramatic effect on the physical and the mechanical properties of the moulded parts, profile extrudates, film, etc. The kind and degree of molecular orientation are largely determined by rheological behaviour of the polymer and the nature of flow in the fabrication process. Furthermore, PP is a semicrystalline material which always be affected by the application of shear stress under the action of shear flow during processing. Simultaneously upon solidification, the generated shear flow will bring a profound effect on the nucleation and crystallization of polypropylene. Crystallization process involves the transportation of molecules from disordered liquid to ordered solid stage. It is a unique phenomenon which occurs due to chain orientation of PP molecular structure. In recent years, many experimental results have confirmed that the orientations of the polypropylene molecules are strongly affected by the flow field acting on the molten PP (Jikan[a] et al., 2010; Jikan, 2010; Rahim[b] et al., 2011).

Other than experimental or practical assistance, rheology can also be a great help to polymer processing in carrying out theoretical analysis of the mechanics of flow of rheologically complex of polymer in various kind of processing equipments. Theoretical model requires rheological models which describe reasonably well the flow behaviour of polymeric materials under consideration. Hence given a flow field of specifically PP, the development of an acceptable rheological model is very important to the success of theoretical study of flow problems. Such a theoretical study should be useful for designing better processing equipment and determining optimal processing conditions. Reviewing the mentioned importance, one should note that behind the complexity of rheology measurement there is significant importance that makes rheological studies unavoidable to polymer scientists and polymer design engineers (Ariff, 2003).

3. Factors affecting the rheological behaviour of polypropylene

Previously, many have reported (Gonzalez et al., 1998; Liang, 2008; Brydson, 1970) on factors affecting the rheological behaviour of PP. Basically, reports usually regulate around four fields of studies. The first field of study involves molecular structure influence on rheological behaviour of PP such as the type of backbone chain, chain branching and chain branching configuration. The second area which is the most likely favoured by researchers is the study on how composition of polymer system or heterogeneity affects rheological behaviour such as those found when PP are blended with other polymers, addition of fillers and other additives are also of interest. The third part involves the study on the dependence of PP rheological behaviour towards fabrication process parameters like pressure, temperature and equipment's geometrical factor. Lastly, the fourth covers the PP theoretical analyses using various rheological models of PP and their implementation in simulation software which regulates around all three fields of studies previously discussed. The extent of these factors will be forwarded in the following sections.

3.1 Molecular structure of polypropylene

PP can be made from the propylene monomer by a process known as Ziegler-Natta catalyzed polymerization or by metallocene catalysis polymerization. PP is a linear hydrocarbon polymer that contains little or no unsaturation in its chain structure. Structurally, it is a vinyl polymer with every other carbon (C) atom in the backbone chain is attached to a methyl (CH₃) group. Compared to polyethylene, PP has some similarities in their characteristics such as swell in solution. The characteristics of PP vary according to the molecular weight and grade. The Ziegler-Natta catalysts have several active sites and accordingly the obtained PP exhibits broad tacticity and molecular weight distributions. It is discovered that by using different types of catalysts and polymerization methods, the molecular configuration can be arranged to produce three types of PP (Jikan, 2010; Karger-Kocsis, 1995).

Mechanical properties, solubility and melt level can be ascertained with knowledge on polymer tacticity. (Andres et al., 2007). The presence of methyl group in PP backbone chain can provide various differing characteristics for PP, depending on the arrangement of methyl group in PP carbon atom, whether in isotactic, syndiotactic or atactic configuration (as shown in Fig. 3). This tacticity will dictate the viscosity of PP that directly contributes to different end properties. Most (90-95%) of commercial PP are isotactic PP produced by Ziegler-Natta catalyst with head-to-tail incorporation of propylene monomer. Isotactic configuration is the most stable structure since the methyl group is arranged at only one side in the PP chain structure. This structure prevents PP from crystallizing in a zig-zag planar shape, but rather in helical crystal structure. With those conditions, the degree of crystallization of isotactic PP can normally reach up to 50% that causes an increase in PP softening temperature which means an increase in the melt viscosity throughout the processing procedure. However, the presence of methyl group attach on the PP backbone are easily oxidized when high processing temperature is utilized and these sites are prone to be chemically attacked by certain chemical agents.

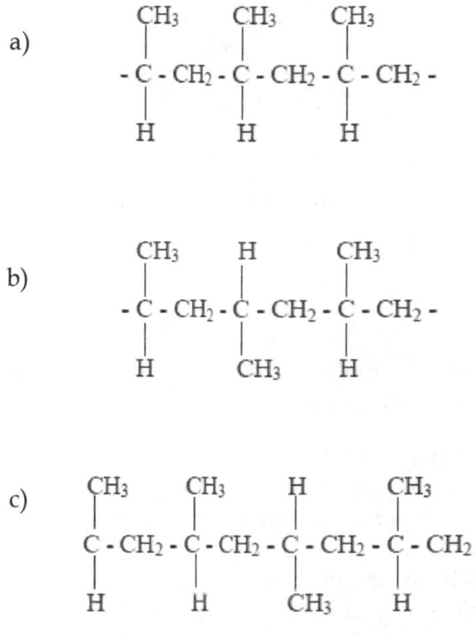

a)
$$CH_3 \quad CH_3 \quad CH_3$$
$$- C - CH_2 - C - CH_2 - C - CH_2 -$$
$$H \quad H \quad H$$

b)
$$CH_3 \quad H \quad CH_3$$
$$- C - CH_2 - C - CH_2 - C - CH_2 -$$
$$H \quad CH_3 \quad H$$

c)
$$CH_3 \quad CH_3 \quad H \quad CH_3$$
$$C - CH_2 - C - CH_2 - C - CH_2 - C - CH_2$$
$$H \quad H \quad CH_3 \quad H$$

Fig. 3. The three basic structures for polypropylene (a) isotactic, (b) syndiotactic and (c) atactic.

However, PP chain configuration is directly related to density which can be classified in many forms such as homopolymer, copolymer with ethylene, or homo- or copolymer blended (which will be discussed later) with ethylene propylene diene monomer (EPDM) rubber or known as thermoplastic polyolefins (TPE-O or TPO, when elastomeric properties appear at higher levels of EPDM). All these components that exist in either PP site chain or partially associate within the PP matrix will definitely influence the molecular weight (MW) and molecular weight distributions (MWD) of the produced PP. It has been revealed in Fig. 4 that the viscosity of PP copolymer (with ethylene monomer) is much lower than that of homopolymer PP obtained using twin screw extruder. The developed viscosity curve has shown to be strongly dependent on the molecular architecture of PP backbone which is well associated with polyethylene group link to its backbone. Higher amount of side group i.e. the methyl side group has fairly raised the disentanglement to the PP polymer chain and caused reduction to the chain mobility. This indicates that the viscosity of PP is virtually dependent on the irrespective molecular chain characteristic (Gonzalez et al., 1996; Andres et al., 2007).

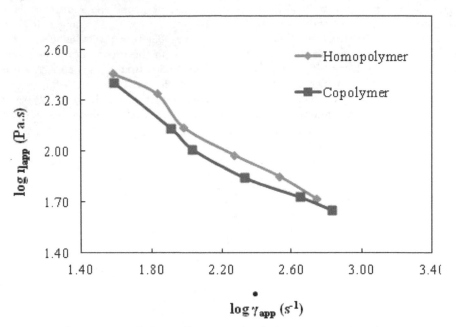

Fig. 4. Viscosity flow curves of PP copolymer (with ethylene monomer) and PP homopolymer obtained using a twin screw extruder

Moreover, the types of manufactured PP differ resulted from their large range of MW. Higher MW usually indicates the existence of longer molecular chains that provide more points of contact or even entanglement between chains. Short branches normally do not affect the viscosity on the molten PP significantly compared to long branching system. Such branching reduces the viscosity of the polymers yet unlike LDPE, PP branch is greater that they can take part in formation of entanglement (Bydson, 1970; Samsudin et al., 2006; Ariff, 2003). Depending on the degree of chain entanglement, it is more likely that a secondary crosslink or known as physical crosslink would form and cause reduction in the mobility of PP molecules. Due to that matter, high shear forces are required to move these PP melts. Gotsis and co-workers has systematically approved that branches existing in polypropylene has increased the zero shear viscosity and increase the polypropylene melt elasticity and improve the melt strength (Gotsis et al., 2004). Hingmann and Marczinki have suggested that the PP shear viscosity increased with chain branching using the Lodge Model (Hingmann and Marczinki, 1994). Meanwhile, Ogawa has finalised that the higher the PP molecular weight, the higher the PP mechanical properties will be (Ogawa, 1994).

In recent years, many researchers (Azizi et al., 2008; Samsudin et al., 2006) are aware and have made considerable interests in dedicating their work in modifying the chain configuration of PP to alter the melt flowability of the material. The most preferred method is by additions of degradation substance such as peroxides. These peroxides are introduced to control the rheological behaviour of PP by lowering the viscosity of the viscous PP melts during processing. Previously, Azizi and colleague has soundly concluded that the addition of high amount dicumyl peroxide has significantly reduced the melt viscosities of PP using either MFI or twin screw extruder. They also found that the addition of peroxide has subsequently lowered the MW and MWD of PP by chain scission reactions which has shortened the PP chain length consequently leads to decreasing values of PP viscosities. The statement is in good agreement with Berzin et al. who also claimed that PP molecular weight distribution are narrowed and high molecular weight species has decreased when the amount of peroxide was increased which enable it to form free radical during thermal decomposition (Berzin et al., 2001).

Another common way to configure the chain achitecture of PP is by thermal, mechanical, oxidative or combinations of these mechanisms. Initially, the identification of these mechanisms would equip manufacturers with good information in selecting PP processing conditions either to promote or to prevent it. It was demonstrated by Gonzalez et. al that during multiple extrusion of PP, MW and viscosity can be reduced with increasing number of extrusion cycles. The chain scission is thought to be provoked by thermomechanical reactions rather than degradations (Gonzalez et al., 1998). On the contrary, PP which contains tertiary hydrogen atoms are most susceptile to oxidation reactions. Other than chain scission, PP also has the tendency to form a crosslinking. The reaction is highly contributed by radical-radical combinations where it literally increase the MW and MWD. Traditionally, intrinsic viscosity measurement were implemented as an indicator of the MW of a PP. However, intrinsic viscosity test have been largely replaced by gel permeation chromatography (GPC) which enables a direct assesment of MW. The occurence of PP crosslinking are commonly found during multiple extrusions with the contributions by various enviromental factors and radiation (Scheirs, 2000).

3.2 Composition or heterogeneity effect of polypropylene

The composition or heterogeneity effects are classified as external factors since all of them are included during the mixing process. One of the factors involving the composition of polymer system includes addition of one or more system which known as blends. Consequently, PP blends have been extensively studied and developed by many researches for the past 20 years either with thermoplastic or elastomeric materials such as ethylene-propylene copolymer (EPR) and ethylene propylene diene monomer (EPDM). Blending is a common useful method to improve PP properties but it is not as simple as adding other polymers into an extruder and mechanically mixing them in the molten state. Typically, morphology and compatibility issue as well as processing difficulties which closely related to flow ability of the blends will rise. The viscosity may increase and decrease depending on the structure of the added block of polymer into the molten PP. Normally the increase in viscosities are caused by the addition of immiscible block like styrene or by the change of bulk properties or the addition of blend component that enhanced the degradation effect in PP blends (Keawwattana, 2002).

The second types of blends that can be produced are miscible blends where the blend may have a higher viscosity than PP and appears to be remarkably elastic. The chemical structure of the miscible block appears to be more important in governing the observed viscosity of the system where it has increased the viscosity of the system. For instance, Song et. al has shown that the rheological behaviour of blend PP and POE (polyethylene-1-octene) is highly depended on the blend ratio of the system. Higher dynamic complex viscosity and loss modulus were observed for the blends with portions of 10 wt% and 20 wt% PEO where miscibility was achieved. Later on, incorporations of higher content of PEO have subsequently reduced the dynamic complex viscosity and loss modulus which attributed to the immiscibility between the two matrices (Song et al., 2008). Besides miscibility, the increase in viscosity of PP blends can also correspond to the effect of crosslinking of the other blend materials. Previously, the rheological behaviour of PP-natural rubber blends was investigated by Thitithammawong and co-workers. They have found that the shear stress and the viscosities of the blends have increased as a result of crosslinking of rubber molecules (Thitithammawong et al., 2007).

On the other hand, filler and reinforcement have always played an important role in modifying PP properties. Fillers incorporated inside PP matrix is purposely to reduce overall cost, improving and controlling process characteristic, density control, dimensional stability and etc. The inclusion of fillers into PP can affect almost all of its properties which also include surface, colour, expansion coefficient, conductivity, permeability, mechanical and rheological properties. Many factors influence the behaviour of fillers in PP melts. The type, compounding method and loading of filler dictate its effectiveness. Fillers can be categorized into two groups, i.e. inert fillers which act as cheapeners and reinforcing fillers that are sometimes used in engineering applications. Among these types of fillers are mineral fillers such as calcium carbonate, talc and kaolin which are most commonly used in PP (Ariffin et al., 2008). The addition of these fillers in PP however, would bring a significant change in the rheological properties such as the viscosity of the base resin. From Fig. 5, it can be seen that the viscosities of PP-kaolin composites have risen as the addition of higher kaolin content measured using a Melt Flow Indexer (MFI) at 210°C. This fact is attributed to the substitution of PP matrix which consists of flexible molecules with more rigid kaolin particles. The ease of melt flow is highly dependent upon the mobility of the molecular chains and force or entanglement holding the molecules together. As the kaolin loading increases, the PP chain mobility are significantly affected by the overloading of kaolin particles in the system which perturb the normal flow and hinder the mobility of chain segments in melt flow, consequently increasing the value of the apparent viscosity (Rahim[a] et al., 2011; Jikan et al., 2009, Rahim, 2010)

Particle shape is another crucial factor in the incorporation of filler in polymer since this affects the polymer characteristics and processing method. Shapes of particles can be categorized as cubic, needle-like (acicular), block, plate or fibre. Spherical particles that flow and disperse well throughout the molten polymer cause the least problem related to stress concentration (Jikan, 2010; Pukanszky, 1995). Needle-like (acicular), fibrous and platy shape fillers can be more difficult to disperse and they can act as stress concentrators, which reduces impact strength (Rothon, 2002). Depending on the filler shape, the addition of fillers at low concentration is able to increase the flow resistance and reduce the built up pressure within the processing equipment. Maiti et al, has mentioned that the apparent viscosity of iPP/CaSiO$_3$ system exhibited lower shear stress compared to pure PP. The reason was claimed to be interrelated with CaSiO$_3$ flake-shape. Same scenario was discovered for PP-kaolin composites where kaolin has similar flaky geometry. The flake-like shape of these filler particles make them able to slide within the PP system during the application of shear forces causing a flow-favouring orientation which subsequently lowered the viscosity of PP matrix (Rahim[a] et al., 2011; Maiti et al., 2002)

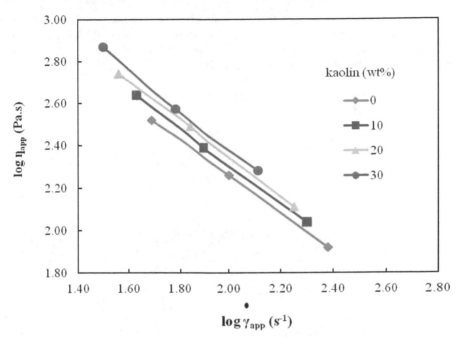

Fig. 5. Viscosity curve of PP-kaolin composites using a Melt Flow Indexer (MFI) at 210°C

Since PP macromolecules do not have polar groups, the homogeneous dispersion of hydrophilic particles in PP is difficult. Therefore, it is necessary to modify the system so that the filler and PP have proper interaction and distribution. To encounter the issue, compatibilizer or coupling agent is usually added to act as a wetting agent between the filler and PP. Among the highly used compatibilizer in PP is maleic anhydride-graft-polypropylene (PPgMA). In the investigation of Liaw et al. the incorporation of PPgMA has reduced the composites viscosity depending on the amount used for PP/clay nanocomposites. To reconfirm the statement, we made an attempt to analyze it with a wide range of shear rate using a capillary rheometer for 20 wt % kaolin loading with the addition of 5 wt% PPgMA and without the addition of PPgMA at 210°C. The claim and our finding is in good approximation where PPgMA has played a role as a flow promoter in PP matrix. The softening effect brought by the maleic anhydride group (MA) content has lowered the system melting temperature. Hence, the decreasing value of the composites viscosities throughout the shear rate range is attributed by the plasticizing effect by the PPgMA in the composites (Liaw et al., 2008).

Other composition factor that affects the processing of PP melt is the inclusion of necessary additives such as plasticizer, lubricant or flow enhancer, environmental stabilising agents and pigment or colorant. The last two usually do not bring significant effect on the flow behaviour of PP melt since they are added in small quantities unless they are able to carry multiple functions which includes plasticizing (similar to PPgMA) and lubricating the compounding system. Nevertheless, the first two additives, i.e. plasticizer and lubricant proved to have a bound effect on the PP melt flow. Due to their ability to dissolve in PP matrix, they are able to space out the PP molecules thus increasing their mobility and consequently reduce the viscosity of the system. Apart from reducing the viscosity, they also tend to reduce the glass transition (T_g) temperature of PP and elastic modulus of the melt (Liaw et al., 2008; Ariffin et al., 2008; Maiti et al., 2002).

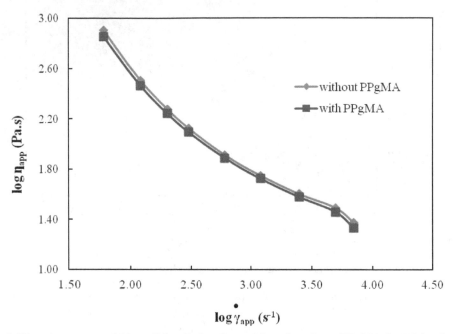

Fig. 6. Viscosity curves of 20 wt % kaolin loading with and without PPgMA for PP-kaolin composites obtained using capillary rheometer at 210°C

3.3 Fabrication process parameters of polypropylene

For the ease of flow, polymer molecules must have enough thermal energy to make it mobile which also associates with having enough space surrounding the molecule which allows it to past other molecules. Thereby, the melting of PP is strongly dependent upon the mobility of its polymer chains. Below T_g, the latter condition is not met, and PP is still in solid state. Above T_g, the magnitude of shear viscosity is totally dependent on the availability of free volume. Whereas at temperature far above the glass transition temperature or the melting point of PP, there is ample free volume available and the temperature dependence of the zero-shear viscosity is determined by energy barriers to motion. A number of researches have come into agreement that the viscosity of PP follows the Arrhenius equation to a good approximation where it is common to observe a decreasing function of viscosity and shear rate that varies considerably (Gupta, 2000).

For Isotactic PP, it is expected that the overall viscosity for the high processing temperature is respectively lower than that of at low processing temperature (Fig. 7). From the molecular level, the flow occurs when PP molecules slide past each other. Whereby, the ease of melt flow depends upon the mobility of PP molecular chains and forces of entanglements holding the molecules together. As the free volume increases with temperature, PP molecules occupy more space due to an increase in the distance between them making it easier to slide among each other and together with the introduction of high shear rate will eventually reduce the PP matrix viscosity significantly (Scheirs, 2000). Similar trend was observed for PP composites system by Rahim[b] et al. and blend system of LDPE/PP blends by Liang and Ness where the viscosities of the compounds are lowered at high processing temperature (Rahim[b] et al., 2011; Liang and Ness, 1998).

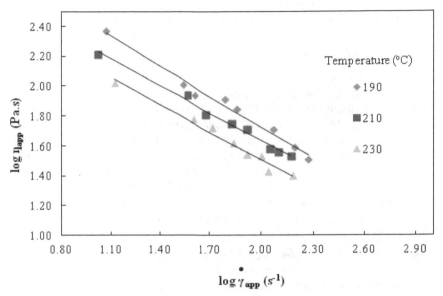

Fig. 7. Viscosity curves of isotactic PP obtained using single screw extruder at various processing temperatures.

In contrast to the effects of temperature and pressure play significant roles in altering the viscosity of polypropylene melt. During polymer processing operations such as extrusion and injection moulding, the applied pressure can reach up to thousands of atmospheres. Large hydrostatic pressure will results in a decrease in the free volume in the melt and accordingly, it should display an increase in viscosity near to polymer T_g. In addition, the reduction of free volume will result in the mobility of polypropylene molecules in the melt to become more restricted. Since viscosity is very dependent towards the distance between each molecule, it is expected that an increasing pressure would increase the viscosity by the way of that a low molecular weight liquids decreases its distance between particle and molecules. Thus, the influence of pressure can be considered quantitatively equivalent to that of temperature but acts the opposite way (Brydson, 1970).

Besides temperature and pressure, in various polymer operations, molten polymer is forced to flow into a wide variety of geometrical configurations; indeed the rheological properties of the polypropylene melt are influenced by the flow geometry (Micheali, 1992; Liang, 2001). Extrusion process have the most critical dependence on flow geometry, where the final product properties as well as appearance is directly related to the geometry or die through where the molten PP is extruded out which will be further discussed in detail in the extrudate swell sections. Experimentally, a non-overlapping rheological experimental data of viscosity curve of different length to diameter (L/D) die ratios can be observed. The results are more obvious for low ratio of L/D. Theoretically, it should form a single curve despite of the different die geometries since the measurements are implemented on the same material. The cause of this problem is highly due to the existence of large error in pressure drop measurement within the measuring instrument. To address the erroneous pressure measurement, Bagley correction procedure was applied and has proven to be successful not just for PP but also for other thermoplastic materials (Rahim, 2010).

4. Viscoelasticity of polypropylene

Viscoelastic properties, as their name implies, display responses towards an applied stress which is a combination of elastic and viscous deformations. The two most commonly used models are those attributed to Maxwell and Kelvin. Kelvin model describes very well the concept of 'creep' (the change in strain at constant stress), whilst the Maxwell model quite reasonably describes 'stress relaxation' (the change in stress at constant strain). In combination, the elements of Maxwell-Kelvin model display both elastic and viscous (namely viscoelastic) characters at deformation rates in between the two extremes (i.e. low and high rates) in varying proportions.   A number of different effects may be noted depending upon the degree of elastic character possessed by a viscoelastic polymer melt (Prentice, 1997). Generally, the relative proportions of each property are highly dependent on the rate of deformation.

It is well known that PP possesses viscoelastic behaviour during polymer processing. Both viscous and elastic component will deform and this would bring about the occurrences of phenomena that are associated with elastic response for PP. During the flowing process, the melt will travelled from a large reservoir (i.e. barrel) to very small die geometry under the application of shear forces. Under normal circumstances, entanglement between molecules prevents the molecules from sliding past each other. When shear forces are introduced, the chain will uncoil and the melt will start to move. On release of the applied shear stress, the chain will recoil and in conjunction they can be pulled back by the restraining forces (i.e. molecular orientation). This theoretically explains the most common effect observed during extrusion process involving extrudate swell and flow instabilities which are going to be deliberated in the next couple of sections.

4.1 Extrudate swell

Extrudate swell is strongly related to elastic recovery of PP at the inlet of a die. Newtonian liquids can also display swelling occurrence, but can only be observed at a very high flow rates. Whereas for PP (which is a non-Newtonian liquid), the swelling will increase with increasing flow rate. When a polymer melt is deformed, either by stretching, shearing or often by a combination of both, the molecules chain are stretched and untangled. In time, the molecules will try to recover their initial shape, by then getting used to their new state of deformation. If the deformation is maintained for a short period of time, the molecules may return to their initial configuration and the shape of the melt is fully restored to its initial shape. It can be said that the molecules remembered their initial configuration. However, if the shearing or stretching goes on for an extended period of time, the molecules cannot recover to their initial arrangement, in essence forgetting their initial positions (Osswald, 1998). Thus, when elastic liquid (such as in PP melt) is extruded from a die or flowed from the exit of a tube, it usually swells to a much greater diameter than that of the die hole. Besides shear rate, the swelling behaviour is highly affected by a number of factors such as temperatures, L/D of the die, shape of the die, filler loading and etc. (Ariffin et al., 2008).

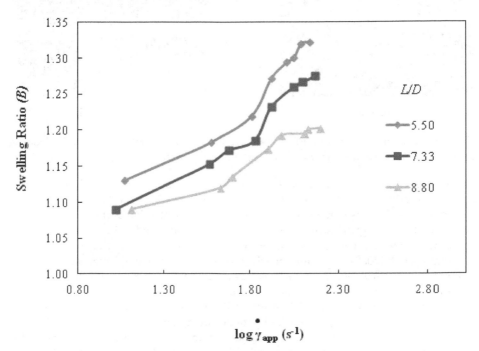

Fig. 8. Swelling ratio of isotactic PP recorded from extrusion process using single screw extruder at various processing temperatures.

One of the affecting parameters for extrudate swell is the influence of die design which is easily illustrated through the die L/D ratio. The extrudate swell phenomenon is indicated by the swelling ratio (B) obtained by dividing of extrudate diameter with the die diameter. For a given rate of flow or shear, the extrudate swell decreases as the L/D ratio of the capillary increases and ultimately attains an equilibrium value as shown in Fig. 8 for pure PP. The root for this condition is the extensional flow at the capillary entrance that imparts greater molecular orientation compared to shear flow within the capillary. Thus, in a long capillary die, some of the molecular orientation imparted to the polymer in the entrance region can relax out in the capillary itself. Also at a constant shear rate, extrudate swell tends to decrease fairly as the temperature is increased since the molecular orientation relaxes faster at higher temperatures (Gupta, 2000). In addition, as the temperature rise, the contributions of viscous component will become more pronounce than that of the elastic component. Furthermore, until it solidifies, the extrudate will sag under the influence of gravity and its diameter will then be reduced. The simplest practice to reduce the extrudate swell is by using a capillary die with a large L/D ratio that will eliminate the effect of the entrance flow on the swelling extrudate.

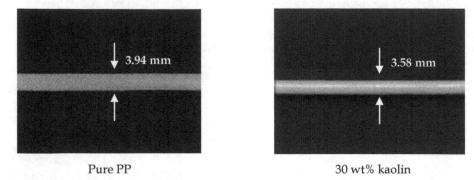

| Pure PP | 30 wt% kaolin |

Fig. 9. Swelled extrudates of pure PP and PP-kaolin composites with 30 wt% filler loading for obtained from extrusion process using capillary rheometer at 200°C

It can be observed from Fig. 9 that PP with 30% kaolin loading had predominantly experienced a reduction of extrudate swell ratio. This behaviour can be due to the fact that the PP chain remains unchanged because of the present of filler particles thus providing more oriented and aligned structures. It is obvious that filled compounds offer greater deformation resistance, due to a reduction in mobility of the polymer chains. This concurs with Muksing et al. (2008) who stated in his research that the layer of fillers create molecular chain barrier in vertical plane towards extrusion direction when leaving the capillary die which limits elastic recovery thus causes reduction in extrudate swell (Liang, 2002). Whereas different scenario can be observed in unfilled PP, an increase in deformation rates in this system is generated by greater molecular realignment and reorientation which reduced relaxation of molecules, which leads to enhanced chain rigidity and shattered melt strength. The addition of rigid particulate filler such as kaolin clearly has similar qualitatively effect on the melt strength of thermoplastics, as noted by Tanaka and White (1980).

4.2 Flow Instabilities

Besides extrudate swell, another common elastic effect occurring in polymer processing is flow instabilities such as melt fracture. Distortion of extrudate is a result of polymer molecules reaching their elastic limit of storing energy, thus causing melt fracture as a way of relieving stress either at the die wall or at the die entrance. Another opinion forwarded by White (1973), is that the extrudate distortion is caused by differential flow-induced molecular orientation between the extrudate skin, holding highly oriented molecules, and the core which has no significant molecular orientation. It is of course, possible that the melt fracture occurs due to a combination of the stress relief theory and the differential flow induced molecular orientation (Shenoy, 1999). In conjunction, PP body consists of long chain molecules and when high pressure is applied; the system would simplify the movement of PP molecules to slide past each other. A rapid and random movement of molecules might cause disordered configuration to the chain that allowed deformations to take place. More likely when stress is removed as the extrudate emerge from the die, the elastic behaviour of the melt flow will struggle to recover predominantly from the elastic memories attained by the large deformation. This elastic property could be seen when extrudate flowing out of the die and it manifest itself in many ways such as the extrudate exhibiting flow instabilities after experiencing extrusion process (Liang and Ness, 1998).

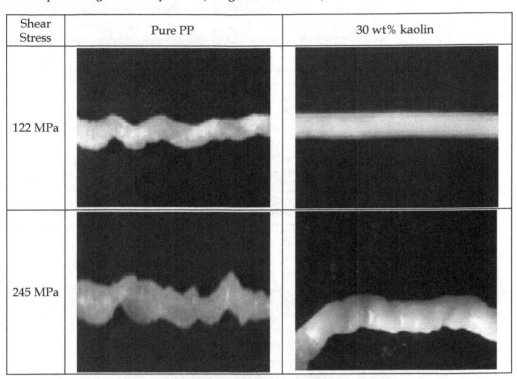

Fig. 10. Flow instabilities of pure PP and PP-kaolin composites with 30 wt% filler loading obtained from extrusion process using capillary rheometer at 165oC for L/D = 4

Since the flow instability is more available for shorter die and at lower die tempereture, an experiment was conducted using a capillary rheometer at 122 MPa and 245 MPa of shear stress for a capillary die with L/D = 4 and at the temperature of 165°C to verify the scene. Based on Fig. 10 for pure PP and 30 wt% kaolin loading for PP-kaolin composites, it has plainly shown that the incorporation of fillers in the composite systems is able to reduce melt flow instability from appearing. This was caused by the presence of a large amount of filler that reduced chain recovery memory by preventing the polymer molecules from returning to their original configuration after stress was withdrawn from the system. Fillers are able to restrict the mobility and deformability of the matrix by introducing mechanical restraint. Whereas, unfilled PP did not have any additional structures that can hold and restrain PP molecules from returning to their original configuration. Furthermore, fillers have the ability to delay melt fracture given that filler particles tend to travel to the surface (skin) of the sample melt thus improving sticking effect between melt and die wall (Gupta, 2000) thus the stick-slip scenario will then be reduced. It clearly elucidates that filler loading highly influences this phenomenon.

Accumulated from the effect of filler loading, preliminary testing has evidenced that low shear stress applied to the sample melt (at low processing temperature of 165°C) was unable to extrude the sample out of the die due to the fact that the system is still in a semisolid state. Nevertheless, different scenario could be seen when the samples were processed using higher shear stress (≥122 MPa). It is believed that the ability of extrudate to emerge from die even at semisolid state is brought about by pressure that surpasses the critical shear stress. When exceeding that value, pressure This led to fluctuation in flow propagation which contributed slip or stick phenomenon (Shore et al., 1997). Extrudate with shear stress of 245 MPa shows further distortion compared to extrudate with shear stress of 122 MPa. This means that flow instability increased when shear stress was increased. The flow instability experienced in this research is commonly identified as melt fracture. Normally, melt fracture is caused by nonlaminar or turbulence flow due to the L/D, type of polymer and processing parameter (Cogswell, 1981). During the extrusion process, the flow direction of PP melt in the barrel was in a form of laminar flow. Nevertheless, it started to experience disturbance when entering the die area. This is due to the fact that the melt in the middle of the die entry started to fracture when high shear stress was applied to the system. This led to fluctuation in flow propagation which contributed towards the flow instability experienced by the extrudate surface.

The effect of temperature plays a vital role in reducing the flow instability of PP system. At a temperature well above melting point, molecules start to enhance their movements. The swift activities of these molecules rise the collisions between them and lead to a molecule's repulsion scenario. Simultaneously, these molecules take up more volume given that the space between them has been increased. At this phase, the molecules are in a tranquil state and some of the elastic memory of the system has faded away due to the relaxation condition. Therefore, when shear is applied to the system (at temperature above melting point) molecular chains could effortlessly align themselves according to the flow path without causing any collisions. In this situation, elastic recovery is relatively low or even absent, consequently decreases the deformation rate which is closely related to the flow instability mechanism. On the contrary, most of the extrudate surfaces exhibit relatively different degree of distortion at low processing temperature. At low temperature, the surface is not as smooth as in the case of samples processed at high temperature. These extrudates in fact have rough, irregular surfaces, loss of glossiness and non-uniform diameter. This is due to the melt fracture that occurs in the elastically deformed polymer, in which the shear stress exceeds the strength of the melt. The extensive slipping and sticking of the polymer layer at the wall of the capillary is also one of the factors that contribute towards extrudate deformation (Georgiou, 2004; Rahim[b] et al, 2011).

Apart from causing extrudate distortion, flow instability for PP may also manifest in other ways such as blisters are formed on the surface of the extrudates and void in the cross section of the extrudates as shown in Fig. 11 obtained at the temperature of 210°C for PP-kaolin composites using a single-screw extruder. The first reason is that as more kaolin was incorporated inside the matrix, the activation energy barrier was reduced due to the particle–molecule interface which led to increasing number of nucleation sites and was able to increase the nucleation rate. Consequently, less force was required to overcome the activation energy barrier, resulting in the formation of more bubbles on the extrudate surface. Whereas for the second reason is due to the melt suddenly goes from high pressure to atmospheric pressure during the emergence of extrudate of the die exit and then undergo a sudden quench in cool water that provided a large temperature gradient (Rahim[a] et al., 2011; Rahim, 2010).

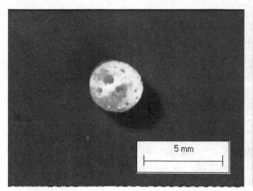
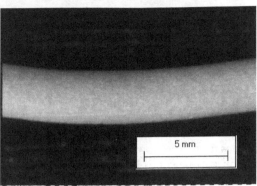

Void in PP-kaolin composites extrudates cross section

Bubble formations on PP-kaolin composites surface

Fig. 11. Flow instabilities for PP-kaolin composites using single screw extruder at 210°C

The extensional flow along the melt will cause an orientation to the polymer chains. Solidification through this type of process condition is favorable for residual stress formation which is one of the factors that can cause shrinkage in many polymer products and in this case, the extrudates. The polymer will freeze the surface in this orientation direction and in the meantime, the meantime, the flow between the solid layer is affected by the temperature gradient resulting in unbalanced cooling where the inner layer cools slowly with respect to the skin layer. As mentioned previously, after exiting the die, the composites will swell first then followed by cooling. Thereby after quenching, the surface layer of the extrudates will solidify first while the inner layer is still in a molten stage. Soon after, as the inner melted layer cools down, it will be attracted to the cooled skin of the extrudate layer. Thus, a void is created in the composites extrudate since the total composites volume reduces with time as the temperature gradient drops (Rahim[a] et al, 2011; Rahim[b] et al., 2011).

5. Conclusion

Overall, the rheological properties of PP are highly dictated by its molecular structure. The chain configuration of the methyl group placement will create different tacticity for PP and subsequently leads to different rheological characteristics. Another contributing factor is the presence of ethylene group such found in some PP copolymers which may increase the flowability of the resin during processing as forwarded in the preceding sections. Besides these internal factors, externally, the rheological behaviour of PP is highly influenced by additives, which are either added to modify the properties of the base compound or for the ease of processability. Incorporation of fillers normally creates higher viscosity of PP melts compared to the pure PP matrix. Whereas, the incorporation of plasticizers, compatibilizers, degradation promoters and incompatible components in the PP blends can correspondingly caused a decreasing trend of viscosity.

Capillary rheometry and extrusion have shown significant importance in measuring rheological behaviour of polymers. As for PP, the above forwarded discussion has proven that its rheological behaviour can be evaluated successfully with both instruments. Our study also proved that both instruments are able to reveal viscoelastic responses of PP melt such as extrudate swell and several types of flow instabilities which cannot be investigated with other types of rheological instrument. The extrudate swell behaviour is caused by a number of factors such as temperature, L/D ratio of the die, flow geometry, filler loading which are eventually connected to the memory effect of the PP melt. Various types of flow instabilities can be observed through capillary rheometry and extrusion process; from periodic extrudate distortion to severe melt fracture. Flow instabilities may also be revealed as loss of glossiness, formation of blistered extrudate surface and even formation of void within the extrudate cross section. These flow instabilities are triggered whenever processing conditions are not optimized and/or exceeded the elastic limit of the PP melt which can be controlled via several approaches such as incorporation of fillers, increasing the die L/D ratio and incorporation of suitable flow promoter.

6. References

Andres, J. A., Pena, B., Benavente, R., Perez, E., Perez, E., Carrada, M. L. (2007). Influence of isotacticity and molecular weight on the properties of metallocenic isotactic polypropylene. *Eur. Polym. J.*, *43, 2357-2370*

Ariff, Z. M. (2003). Melt rheology and injection moulding of acrylonitrile-butadiene-styrene (ABS) using a capillary rheometer. *Phd Thesis*, Universiti Sains Malaysia

Ariffin, A., Mansor, A.S., Ariff, Z. M., Jikan, S.S., Ishak, Z.A.M. (2008). Effect of filler treatments on rheological behaviour of calcium carbonate and talc-filled polypropylene hybrid composites *J. Appl. Polym. Sci*, *108, 3901-3916*

Ariffin, A., Jikan, S. S., Samsudin, M. S. F., Ariff, Z. M., Ishak, Z. A. M. (2006). Melt elasticity phenomenon of multicomponent (talc and calcium carbonate) filled polypropylene. *J. Reinf. Plas. Compos.*, *25, 913-923*

Ariffin, A., Ariff, Z. M., Jikan, S. S. (2011). Evaluation on extrudate swell and melt fracture of polypropylene/kaolin composites at high shear stress. *J. Reinf. Plas. Compos.*, *30, 609-619*

Azizi, H., Ghasemi, I., Karrabi, M. (2008). Controled-peroxide degradation of polypropylene: rheological properties and prediction of MWD from rheological data. *Polym.Test.*, *27, 548-554*

Berzin, F., Vergnes, B., Delamare, L. (2001). Rheological behavior of controlled-rheology polypropyles obtained by peroxide-promoted degradation during extrusion: Comparison between homopolymer and copolymer. *J. Appl. Polym. Sci*, *80, 1243-1252*

Brydson J.A. (1970). *Flow Properties of Polymer Melts*. London; Iliffe Books.

Cogswell, N. (1981) *Polymer melt rheology: A guide for industrial practice*. Woodhead Publishing Limited.

Georgiou, G. (2004) Stick-Slip Instability. Polymer Processing Instabilities: Control and Understanding. Marcel Dekker, New York.

Gonzales, V. A., Velazquez, G. N., Sanchez, J. L. A. (1998). Polypropylene chain scissions and molecular weight changes in multiple extrusions. *Polym. Degrad. Stab.*, *60, 33-42*

Gotsis, A. D., Zeevenhoven, B. L. F., Tsenoglou, C. (2004). Effect of long branches on the rheological behaviour of polypropylene *J. Rheo. 48, 895-914*

Gupta, R.K. (2000). *Polymer and Composite Rheology*. Marcel Dekker Inc. New York

Han C. D. (1976). *Rheology in Polymer Processing*. Academic Press. New York

Hingmann, R., Marczinke, B. L. (1994). Shear and elongational flow properties of polypropylene melts. *J. Rheo. 39, 573-588*

Jikan, S. S., Samsudin, M. S. F., Ariff, Z. M., Ishak, Z. A. M., Ariffin, A. (2009). Relationship of rheological study with morphological characteristics of multicomponent (talc and calcium carbonate) filled polypropylene hybrid composites. *J. Reinf. Plast. Compos. 28, 2577-2587*

Jikan S. S. (2010). Evaluation on flow behaviour of polypropylene/kaolin composites at high shear stress. *Phd Thesis*, Universiti Sains Malaysia.

Jikan, S. S[a]., Ariff, Z. M., Ariffin, A. (2010). Influence of filler content and processing parameter on the crystallization behaviour of PP/kaolin composites. *J. Therm. Anal.Calorm., 102, 1011-1017*

Karger-Kocsis J. (1995) *Polypropylene: structure and morphology.* Chapman and Hall, Cambridge.

Keawwattana, W. (2002). Phase behaviour, crystallization and morphological development in blends of polypropylene (PP) isomers and poly(ethylene-octane) copolymer. *Phd Thesis.* University of Akron.

Liang, J. Z. (2001). Pressure effect of viscosity for polymer fluids in die flow. *Polym., 42, 3709-3712*

Liang J. Z., Ness J. N. (1998). The melt die-swell behaviour during capillary extrusion of LDPE/PP blends. *Polym. Test., 17, 179-189*

Liang, J. Z. (2008) Effects of extrusion conditions on die-swell behavior of polypropylene/diatomite composite melts. *Polym. Test., 27, 936-940.*

Liang, J. Z. (2002) The Melt Elastic Behaviour of Polypropylene/Glass Bead Composites in Capillary Flow. *Polym. Test, 21, 927-931.*

Liaw, W. C., Huang, P. C., Chen, C. S., Lo, L. C., Chang J. L. (2008). PPgMA/APTS compound coupling compatibilizer in PP/Clay hybrid nanocomposites. *J. Appl. Polym. Sci.,109, 1871-1880*

Maiti, S. N., Singth, G., Ibrahim, M. N. (2003). Rheological properties of calcium silicate-filled isotactic polypropylene. *J. Apply. Polym. Sci., 8, 1511-1218*

Muksing, N., Nithitanakul, M., Grady, B. P., Magaraphan, R. (2008). Melt rheology and extrudate swell of organobentonite-filled polypropylene nanocomposites. *Polym. Test., 27, 470-479*

Micheali, W. (1992). *Extrusion dies for plastic and rubber: design and engineering.* Munich. Hanser Publishers.

Ogawa, T. 1992. Effect of molecular weight on mechanical properties of polypropylene. *J. Appl. Polym. Sci, 31, 1151-1154*

Osswald, T.A. (1998) *Polymer processing fundamentals.* Munich. Carls Hanser Verlag.

Prentice, P. (1997). *Rheology and its role in plastics processing, Vol 7*, Rapra Technology Limited, United Kingdom.

Rahim N. A. A. (2010). Flow behaviour and viscoelastic properties of polypropylene-kaolin composites. *MSc Thesis*, Universiti Sains Malaysia.

Rahim,[a] N. A. A., Ariff, Z. M., Ariffin, A. and Jikan, S. S. (2011). A study on the effect of filler loading on flow and viscoelastic behavior of polypropylene/kaolin composites. *J. Appl. Polym. Sci., 119, 73-83*

Rahim[b], N. A. A., Ariff, Z. M., Ariffin, A. (2011). Flow behaviour and viscoelasticity of polypropylene-kaolin extruded composites at different temperatures. *Pertanika J. Sci. & Technol. 19, 383-388*

Rothon, R.N. (2003) *Particulate-filled polymer composites*. Rapra Technology Limited, United Kingdom.

Samsudin, M. S. F., Ishak, Z. A. M., Jikan, S. S., Ariff, Z. M., Ariffin, A. (2006). Effect of filler treatment on rheological behaviour of calcium carbonate and talc-filled polypropylene hybrid composites. *J. Appl. Polym. Sci*, *102, 5421-5426*.

Scheirs, J. (2000). *Compositional and Failure Analysis of Polymers*. England: John Wiley & Sons, Ltd.

Song, N., Zhu, L., Yan, X., Xu, Y., Xu, X. (2008). Effect of blend composition on the rheology property of polypropylene/poly (ethylene-1-octene) blends. *J. Mater. Sci.*, *43, 3218-3222*.

Shenoy, A.V., Saini, D.R. (1996) *Thermoplastic melt rheology and processing*. Marcel Dekker, Inc., New York.

Tanaka, H., White, J. L. (1980) Experimental investigations of shear and elongational flow properties of polystyrene melts reinforced with calcium carbonate,titanium dioxide and carbon black. *Polym Eng. Sci, 20, 946-956*.

Thitithammawong, A., Nakason, C., Sahakaro, K., Noordermer., J. (2007). Effect of different type of peroxide on rheological, mechanical, and morphological properties of thermoplastic based vulcanizates based on natural rubber/ polypropylene blends. *Polym. Test., 26, 537-546*

Wen, S. H., Liu, T. J., Tsou, J. D. (1994). Three-dimensional finite element analysis of polymeric fluid flow in an extrusion die. Part I: Entrance effect. *Polym. Eng. Sci., 14, 212-222*

White, J. L. (1973). Critique on flow patterns in polymer fluids at the entrance of a die and instabilities leading to extrudate distortion. *Appl. Polym. Symp, 20, 155-174*.

Yu, Y. W., Wu, P. V., Liu, T. J. (1997). Validity of one-dimensional equation governing extrusion die flow. *AlChe Journal, 43, 3117-3120*

6

Decomposition of Artificial
Litter Made of Polypropylene

M. Szanser
Polish Academy of Sciences,
Centre for Ecological Research, Łomianki,
Poland

1. Introduction

Environmental pollution with plastic, especially polyolefins increased during the last decades. Due to recalcitrance of these substrates the studies on procedures speeding up the degradation and biological decomposition of polyolefins are carried intensively. The data concerning polypropylene (PP) disappearance rate are scarce comparing to the decomposition of polyethylene (PE) products (Ammala et al., 2011, Arutchelvi et al., 2008, Eubeler et al., 2010). It has been shown that pretreated PP disappear more quickly than poly(propylene-co-ethylene) (CPP) and PE polymers (Arutchelvi et al., 2008, Meligi et al., 1995, Sivan, 2011). However slower degradability of PP comparing to other plastics has been reported (Yang et al., 2005). It is known that UV radiation and weathering processes can speed up biodegradation processes of plastics in later stages of disappearance. Some studies of the PP and PE biodegradation showed that microorganisms, both bacteria and fungi, colonised the pre-treated or composted plastic substrata to greater extent comparing to control treatments (Ammala et al., 2011, Arutchelvi et al., 2008, Eubeler et al., 2010, Grunz et al., 1999, Meligi et al., 1995). There is a little information about the biodegradability of PP in the natural environment and impact of fresh organic matter input on PP decomposition. It seems that the natural input of nitrogen-rich organic matter both of plant and invertebrate origin might accelerate microbial activity and in consequence speed up the decomposition processes of recalcitrant materials (Crow et al., 2009, Dekker et al., 2005, Griffiths, 1994 Prévost-Bouré et al., 2010, Szanser, 2000).

The aim of the study was to compare the decomposition rate of the PP and natural plant litter (NPL). The hypothesis was that the decomposition of artificial litter made of PP will proceed effectively as in the case of natural litters due to additive effects of the input of plant and invertebrate remains into the litter.

2. Materials and methods

2.1 Study site and experimental design

The study was carried out in a permanent meadow of the type Arrhenatheretalia situated in the buffer zone of Kampinos National Park (52°15'30" N and 20°17' E, east-central Poland).

An experiment was conducted to compare the decomposition pattern of natural plant litters (NPL) and of artificial litter composed of natural not dyed PP agricultural string. The experimental meadow of an area of 190 x 10 m had 182 microplots (0.5 x 0.5m area and 0.15 m depth). Five natural and one artificial litter treatments in 32 and 22 plot replicates respectively according to RCB design were applied (Pearce, 1983). Eight litter containers per plot were placed. Natural litter was obtained from meadow plants, both grasses and weeds cut in August 2001 which had to simulate the input of decaying plants to soil. The same amount of PP and NPL (9 g dry wt.) irrespective of the number of plant species was exposed in modified litter containers. Experimental plots were filled with sand mixed with loam to the depth of 15 cm. Such type of simplified and uniform substrate, among them, enabled assessing the input of organic matter morphous particles in exposed litter and in underlying substrate.

Litter samples were mounted under stainless steel wires stuck into the substrate. In that way a more compact structure of exposed material which adhered to the substratum was obtained, providing fauna with better access to the litter without covering containers by a net on the top. NPL were either monospecific (I: *Dactylis glomerata*; II: *Festuca rubra* and III: *Trifolium pratense*) or were species mixtures (IV: mixture of three species I, II and III; V: mixture of twelve species, IV and nine other meadow plants). In the latter case, the following species were used: grasses: brome grass (*Bromus inermis*), meadow foxtail (*Alopecurus pratensis*), perennial ryegrass (*Lolium perenne*), oat grass (*Arrhenatherum elatius*), cocksfoot (*Dactylis glomerata*), and red fescue (*Festuca rubra*); and herbs: small plantain (*Plantago lanceolata*), common chicory (*Cichorium intybus*), red clover (*Trifolium pratense*), milfoil (*Achillea millefolium*), carrot (*Daucus carota*), and common silverweed (*Potentilla anserina*).

2.2 Sample collecting and processing

The experiment started on 24-25 March 2002. Samples were taken on 27 September 2002, 11 May 2004, 9 September 2004, i.e., 6, 26, and 30 months after litter exposure.

Litter and soil samples were taken per 5-11 in every sampling occasion. Soil samples for assessments of plant and invertebrate organic matter input were taken to the depth of 1 cm with a soil corer 100 cm^2 in area.

2.3 Plant and invertebrate materials

2.3.1 Litter mass loss

Litter mass loss was determined using the gravimetric method by drying at 65°C.

2.3.2 Input of wind-borne plant matter and invertebrate remnants into litter and underlying substrate

The input of the remnants of invertebrate origin (exuviae, cocoons, other remants) was assessed in the litter and soil layer 0-1 cm underneath the litter by hand-sorting and using the stereoscope microscopy.

2.4 Statistical analysis

Statistical analyses of results (one-way *ANOVA* and regression) were performed using Statistica 8.0. software (StatSoft, Inc. (2007). Natural litter and remnants' data had no normal distribution. Only total data for PP litter and remnants had normal distribution. Differences in mass of litter and remnants between treatments were assessed using non-parametric Wilcoxon Sum-of-Ranks (Mann-Whitney) test for comparing two unmatched samples. The regression between masses of remaining litter and remnants were assessed using the Kendall tau Rank Correlation test and linear regression analysis for natural and polypropylene treatments, respectively.

3. Results

3.1 Litter decomposition

The highest values of the NPL mass loss were recorded in the initial period, while the loss of PP litter was the slowest (Table 1). In the first 6 months following exposure, the loss of dry weight mass of litter reached 59.8 and 28.1% respectively for NPL and PP treatments (Table 1). Later, further losses of litter, during over two years period, were observed being 70.53 and 72.45% for NPL and PP treatments respectively. In the final period, i.e., between the 26th and 30th months of the experiment, the increases of both types litter mass were larger than its losses. This could be explained by the organic matter (of plant and animal origin) input and colonisation of the litter by invertebrates and microorganisms during long period of the exposure (2.5 years).

Months since beginning of the experiment		Litter type		Significance of differences between treatments
		NPL	PP	
6	g dry weight m^{-2}	362.28 (7.22)	646.78 (24.69)	n = 55, W = 502.5, p<=0.00000
	%	40.25	71.86	
26	g dry weight m^{-2}	265.26 (9.98)	337.98 (28.61)	n = 35, W = 106.5, p<=0.4229
	%	29.47	37.55	
30	g dry weight m^{-2}	293.78 (14.23)	360.88 (58.91)	n = 35, W = 111.5, p<=0.1573
	%	32.64	40.1	

Table 1. Mass (g dry weight m^{-2}) of remaining and percent (%) of initial mass of exposed natural plant (NPL) and polypropylene (PP) litter during the course of the experiment. Significance of differences between treatments are assessed by non-parametric Wilcoxon Sum-of-Ranks (Mann-Whitney) test. All data from five natural litter treatments (I: *Dactylis glomerata*; II: *Festuca rubra* and III: *Trifolium pratense*), IV: mixture of three species I, II and III; V: mixture of twelve species, IV and nine other meadow plants) were analyzed with the exception of 26th month where only data of treatments I and III were used. Standard errors in parentheses.

Differences between NPL(data taken together) and PP treatments in the litter decomposition rate resulted from litter origin. Remaining mass of artificial litter (PP) was higher in every sampling time by 43.9%, 27.4% and 22.8% comparing to natural (NPL) ones respectively after 6, 26 and 30 months of the experiment but the differences were significant only for the first sampling occasion (Table 1). Interestingly slight increase of litter mass occurred in PP treatment similarly as in natural litter at the end of the experiment.

3.2 Input of plant and invertebrate remnants

There was considerably large input of wind-borne plant material into exposed litter. It was still low after 6 months of the experiment and was particularly high by the 26th month. There was further slight increase of plant remnants input between 26th and 30th month of the experiment. PP litter had significantly higher input of wind-borne plant matter comparing to NPL only after 6 months of the experiment (Table 2). Later the differences between treatments were not significant. The values of wind-borne plant materials were 43.4 and 39.6 g dry wt. m^{-2} for NPL and PP treatments respectively at the end of the experiment. This input constituted 4.82 and 4.4 % of the initial mass of natural and artificial litter, respectively. Much smaller was the mass of invertebrate origin (exuviae, cocoons, other remnants) found in both treatments but the differences between treatments were not significant (Table 3). The mass of invertebrate remnants amounted to 2.0 and 1.2 g dry wt. m^{-2} for NPL and PP treatments respectively at the end of the experiment.

Months since beginning of the experiment	Litter type		Significance of differences between treatments
	NPL	PP	
6	3.71 (0.37)	9.31 (1.99)	n = 60, W = 305, p <= 0.9921
26	36.77 (9.26)	14.14 (2.67)	n = 24, W = 69, p <=0.7139
30	43.40 (5.72)	39.60 (15.52)	n = 51, W = 138, p <= 0.6089

Table 2. Mass of wind-borne plant remnants (g dry weight m^{-2}) in natural plant (NPL) and polypropylene (PP) litter during the course of the experiment. Differences between treatments are assessed by non-parametric Wilcoxon Sum-of-Ranks (Mann-Whitney) test. All data from five natural litter treatments (I: *Dactylis glomerata*; II: *Festuca rubra* and III: *Trifolium pratense*), IV: mixture of three species I, II and III; V: mixture of twelve species, IV and nine other meadow plants) were analyzed with the exception of 26th month where only data of treatments I and III were used. Standard errors in parentheses.

Months since beginning of the experiment	Litter type		Significance of differences between treatments
	NPL	PP	
6	1.48 (0.23)	2.49 (0.50)	n = 60, W = 305, p <= 0.9921
26	0.85 (0.33)	0.58 (0.33)	n = 24, W = 75, p <= 0.9734
30	2.00 (0.36)	1.17 (0.33)	n = 51, , W = 156, p <= 0.9883

Table 3. Mass of invertebrate remnants (g dry weight m^{-2}) in NPL and PP litter during the course of the experiment. Differences between treatments are assessed by non-parametric Wilcoxon Sum-of-Ranks (Mann-Whitney) test. All data from five natural litter treatments (I: *Dactylis glomerata*; II: *Festuca rubra* and III: *Trifolium pratense*), IV: mixture of three species I, II and III; V: mixture of twelve species, IV and nine other meadow plants) were analyzed with the exception of 26th month where only data of treatments I and III were used. Standard errors in parentheses.

Summarized data of both plant and invertebrate material revealed that significantly higher was the mass of new organic material in PP comparing to natural treatments after 6 months of the experiment (Table 4). Later the differences between treatments were not significant. The values of total organic matter input were 45.4 and 40.8 g dry wt. m^{-2} for natural and PP treatments respectively at the end of the experiment.

The negative correlations between mass of remaining litter and input of both types of remnants were found for NPL (r = -0.456, Z = -7.230, P <0.00000, n = 115) and PP (r = -0.415, r^2 = 0.172, F = 3.743, P <0.0689, n = 20) treatments taking the entire study period data.

Months since beginning of the experiment	Litter type		Significance of differences between treatments
	NPL	PP	
6	5.19 (0.42)	11.8 (2.08)	n = 60, W = 305, p <= 0.9921
26	37.62 (9.32)	14.73 (2.66)	n = 24, W = 69, p <=0.7139
30	45.40 (5.73)	40.78 (15.66)	n = 51, W = 138, p <= 0.6089

Table 4. Sum of plant and invertebrate remnants (g dry weight m^{-2}) in NPL and PP litter during the course of the experiment. Differences between treatments are assessed by non-parametric Wilcoxon Sum-of-Ranks (Mann-Whitney) test. All data from five natural litter treatments (I: *Dactylis glomerata*; II: *Festuca rubra* and III: *Trifolium pratense*), IV: mixture of three species I, II and III; V: mixture of twelve species, IV and nine other meadow plants) were analyzed with the exception of 26th month where only data of treatments I and III were used. Standard errors in parentheses.

4. Discussion

The meteorological data indicated that climatic conditions in the study period were unfavourable for soil organisms (Szanser et al., 2011). Artificial litter (PP) was drier comparing to natural litters (NPL) and retained a maximum of 5% moisture even during rainfall, while its underlying substrate was similarly wet as that under natural litters (Szanser et al., 2011). Lower mass loss of PP comparing to NPL was coincided with significantly lower numbers of fungi and bacteria in the PP litter and its respiration by 3.4, 44.8 and 61.4 times respectively comparing to natural treatments (Górska unpubl., Szanser et al., 2011). Interestingly number of fungi was the parameter the least differentiating between PP and other treatments. It may signify that the decomposition of PP was of fungal nature. Nevertheless it was found that almost 60 and 70% of the exposed PP and natural litters degraded by 26th month of the experiment. It seems that degradation of PP was quite effective as far as there were no significant differences between applied treatments at the end of the experiment. Thirty months after litter exposure, the input of wind-borne plant material was similar in both treatments. Considerably smaller was the mass of invertebrate origin (exuviae, cocoons, other remnants in both treatments. Similar input of invertebrate organic matter found in later stages of the experiment is corroborated by the lack of significant differences between PP and natural litters in invertebrate macrofauna penetration (Kajak, Szanser, unpubl.). Similar values of the input of plant and invertebrate remnants into natural litter and underlying substrate were found in other agricultural and meadow environments (Szanser,

2003, Szanser, unpubl.). It seems that input of organic matter into PP had some effect on its biodegradation. It should be pointed out that the absorption of different ions (Na^+, $N-NH_4^+$, K^+, Mg^{2+}, Cl^-, $N-NO_3^-$, $S-SO_4^{2-}$) by exposed PP from aerosol-gaseous input, although not measured in this study, can have additional impact on speeding up PP biodegradation. It was found that absorption of elements by polymers used as "artificial leaves" can be quite high and increase with surface area of the exposed plastic (Kram, 2005, Stachurski & Zimka 2000). It seems that the aerosol-gaseous input of elements together with input of new organic matter and further development of microorganisms in fibrous PP resulted in efficient decomposition of the PP. Observed slight increase of polymer mass towards the end of the experiment simultaneously as in natural litter can be explained by the organic matter input and development of microbial communities. The mechanisms involved in PP biodegradation are corroborated by data obtained from experimental underlying substrates. There were no significant differences in microbial activity (soil respiration, microbial biomass and numbers of bacteria and fungi) between PP and NPL treatments in underlying substrate during the third year of the experiment (Górska, Szanser, unpubl.). On the opposite algal biomass and production were by 36% and 39.7% respectively higher under PP comparing to NPL treatments for entire study time (Sieminiak, unpubl.). The increase of carbon content under the PP was higher by 47% than in natural treatments at the end of the experiment (Kusińska & Szanser unpubl.). It should be pointed out that the paper presents results obtained from long-time field experiment while most of the research on degrading the polymers are short time studies.

5. Conclusions

In general, the results suggest that (1) the long term decomposition of artificial litter (PP) proceeded efficiently but was still quite low comparing to natural treatments; (2) the input of plant and invertebrate remains into the PP can be considerably large and may have an additive effect on its decomposition; (3) it seems that longtime decomposition of natural PP may be quite effective comparing to plant litters and (4) the presence of PP did not inhibit severely soil biota activity and organic matter accumulation in soil during the experiment.

6. Acknowledgments

This research was supported by a grant from the Polish State Committee for Scientific Research, project P04 F 03820.

7. References

Ammala, A., Bateman, S., Dean, K., Petinakis, E., Sangwan, P., Wong, S., Yuan, Q., Yu, L., Patrick, C., and Leong, K. H. (2011). An overview of degradable and biodegradable polyolefins. *Progress in Polymer Science* Vol. 36, No. 8, (August 2011), pp. 1015-1049, ISSN: 0079-6700.

Arutcholvi, J., Sudhakar, M., Arkatkar, A., Doble, M., Bhaduri, S., and Uppara, P. V. (2008). Biodegradation of polyethylene and polypropylene. *Indian Journal of Biotechnology*, Vol. 7, No. 1, (January 2008), pp. 9-22, ISSN: 0972-5849.

Chemidlin Prévost-Bouré, N., Soudani, K., Damesin, C., Berveiller, D., Lata, J.-C., and Dufrêne, E. (2010). Increase in aboveground fresh litter quantity over-stimulates soil respiration in a temperate deciduous forest. *Applied Soil Ecology*, Vol. 46, No. 1, (September 2010), pp. 26-34, ISSN: 0929-1393.

Crow, S. E., Lajtha, K., Bowden, R. D., Yano, Y., Brant, J. B., Caldwell, B. A., and Sulzman, E. W. (2009). Increased coniferous needle inputs accelerate decomposition of soil carbon in an old-growth forest. *Forest Ecology and Management*, Vol. 258, No 10, (October 2009), pp. 2224-2232 ISSN: 0378-1127.

Dekker, S. C., Scheu, S., Schröter D., Setälä, H., Szanser, M. and Traas, T. P. (2005). Towards a new generation of dynamical soil decomposer food web models. Pages 590 *in* P. C. De Ruiter, V. Wolters, and J. C. Moore, editors. *Dynamic Food Webs -multispecies assemblages, ecosystem development and environmental change*, (No information 2006) pp. 258-269 Elsevier, Amsterdam, ISBN: 978-0-12-088458-2.

Eubeler, J. P., Bernhard, M., and Knepper, T. P. (2010). Environmental biodegradation of synthetic polymers II. Biodegradation of different polymer groups. *TrAC Trends in Analytical Chemistry* , Vol. 29,No 1, (January 2010), pp. 84-100, ISSN: 0165-9936.

Griffiths B. S. 1994 - Microbial-feeding nematodes and protozoa in soil: Their effects on microbial activity and nitrogen mineralization in decomposition hotspots and the rhizosphere - *Plant and Soil*, Vol. 164, No 1, (January 1994), pp. 25-33, ISSN: 0032-079X.

Kram K.J. (2005). Bulk precipitation and aerosol-gaseous input of elements in the forested area of coastal region (Pomerania, north Poland). *Polish Journal of Ecology*, Vol. 53, No. 2, pp. 261–268, (June 2005), ISSN 1505-2249.

Meligi, G., Yoshii, F., Sasaki. T., Makuuchi, K., Rabie, A. M. and Nishimoto, S-I., 1995. Comparison of the degradability of irradiated polypropylene and poly(propylene-co-ethylene) in the natural environment, *Polymer Degradation and Stability*, Vol. 49, No. 2, (No information 1995), pp 323-327, ISSN: 0141-3910.

Grunz, A., Dayss, E., and Leps, G. (1999). Biomass covering of plastics substrates depending on plasma treatment. *Surface and Coatings Technology* Vol. 116-119, (September 1999), pp. 831-835, ISSN: 0257-8972.

Pearce, S.C. (1983). *The Agricultural Field Experiment: a Statistical Examination of Theory and Practice*. New York: John Wiley & Sons, (No information 1983), pp. 335, ISBN-10: 0471105112.

Sivan, A. (2011). New perspectives in plastic biodegradation. *Current Opinion in Biotechnology* , Vol. 22, No. 3, (June 2011), pp. 422-426, ISSN: 0958-1669.

Stachurski, A., and Zimka, J. R. (2000). Atmospheric input of elements to forest ecosystems: a method of estimation using artificial foliage placed above rain collectors. *Environmental Pollution*, Vol. 110, No. 2, (November 2000), pp. 345-356, ISSN: 0269-7491.

STATISTICA (data analysis software system), version 8.0. www.statsoft.com).

Szanser M. 2000 - Effect of macroarthropods patrolling soil surface on decomposition rate of grass litter (*Dactylis glomerata*) in a field experiment – Polish Journal of Ecology Vol. 48, No. 4, (October 2000), pp. 283-297, ISSN 1505-2249.

Szanser, M. 2003 - The effect of shelterbelts on litter decomposition and fauna of adjacent fields: In situ experiment - Polish Journal of Ecology, Vol.51, No.3, (September 2003), pp.309-321, ISSN 1505-2249.

Szanser, M., Ilieva-Makulec, K., Kajak, A., Kusińska, A., Kisiel, M., Olejniczak, I., Russel, S., Sieminiak, D., Wojewoda, D. (2011). Impact of litter species diversity on decomposition processes and communities of soil organisms - *Soil Biology and Biochemistry*, Vol. 43, No. 1, (January 2011), pp. 9-19, ISSN: 0038-0717.

Yang, H.-S., Yoon, J.-S., and Kim, M.-N. (2005). Dependence of biodegradability of plastics in compost on the shape of specimens. Polymer Degradation and Stability, Vol. 87, No. 1, (December 2004), pp. 131-135 ISSN: 0141-3910.

7

Modified Atmosphere Packaging for Perishable Plant Products

Leonora M. Mattos[1], Celso L. Moretti[1] and Marcos D. Ferreira[2]
[1]*Embrapa Vegetables,*
[2]*Embrapa Instrumentation*
Brazil

1. Introduction

Packaging perishable plant products is one of the more important steps in the long and complicated journey from grower to consumer. Millions of different types of packages are used for produce around the world and the number continues to increase as the industry introduces new packaging materials and concepts. Packing and packaging materials contribute a significant cost to the produce industry; therefore it is important that packers, shippers, buyers, and consumers have a clear understanding of the wide range of packaging options available (Boyette et al., 1996). This fact chapter describes some of the many types of packaging materials, including their functions, uses, and limitations. Within packaging plastics for plant products, if commodity and film permeability characteristics are properly matched, an appropriate atmosphere can evolve passively through consumption of O_2 and production of CO_2 during respiration (Mir & Beaudry, 2002). Gas exchange and respiration rate through the package material are the processes involved in creating a modified atmosphere inside a package that will extend shelf life of fresh fruits and vegetables. The major methods for measuring respiration rates, along with their advantages and limitations are discussed. Modified atmosphere technologies have great potential in a wide range of applications in plant products. The usual methods of respiration rate determination can be the static system, the flowing system and the permeable system (Fonseca et al., 2002). The respiration rate of fresh produce can be expressed as O_2 consumption rate and/or CO_2 production rate. Factors affecting the respiration rate and respiratory quotient are outlined, stressing the importance of temperature, O_2 and CO_2 concentrations, and storage time (Kader et al., 1989). Modified atmosphere packaging should always be considered as a supplement to proper temperature and relative humidity management. The differences between beneficial and harmful concentrations of oxygen and carbon dioxide for each kind of produce are relatively small, so great care must be taken when using these technologies. Temperature has been identified as the most important external factor influencing respiration (Tano et al., 2007). The internal factors affecting respiration are the type and maturity stage of the commodity. Vegetables include a great diversity of plant organs such as fruits, roots, tubers, seeds, bulbs, sprouts, leaves and stems that have different metabolic activities and consequently different respiration rates. Different varieties of the same product exhibit specific respiration rates. The success of modified atmosphere packaging greatly depends on the accuracy of the predictive respiration rate (Kader, 2002). The main objective of this chapter is to present different packaging materials using modified atmosphere for perishable plant products, focusing particularly on aspects of the respiration process, usual methods of measuring respiration rates and factors can be affect the respiration rate.

2. Modified atmosphere packaging

Modified atmosphere packaging (MAP) of fresh fruits and vegetables is based on modifying the levels of O_2 and CO_2 in the atmosphere produced inside a package sealed with some type of polymer film. It is desirable that the natural interaction that occurs between the respiration of the product and the packaging generates an atmosphere with low levels of O_2 and / or a high concentration of CO_2. The growth of organisms that cause decay is thereby reduced and the life of the product is thus extended. Additionally, the desired atmosphere can reduce the respiration rate, and ethylene production, physiological changes. For example, it can inhibit chemical, enzymatic and microbiological mechanisms associated with the decay of fresh products, thus avoiding the use of other chemical or thermal process such as freezing, dehydration, and sterilization (Kader et al. 1989; Gorris & Tauscher, 1999; Saltveit, 1997; Fonseca et al., 2002).

The use of modified and controlled atmospheres has grown over the past 50 years, contributing significantly to extend the postharvest life and maintain the quality of various fruits and vegetables. So that changing the atmosphere occurs must have a combination of factors that influence the permeability of the product packaging and respiration in order to achieve an atmosphere of great balance for the conservation of the product. This balance is achieved when the respiration of the product consumes the same amount of O_2 entering the packaging and the production of CO_2 by respiration is equal to the amount that leaves the packaging (Day, 1993).

The first studies on modified atmospheres used reduced levels of O_2 in apple packaging in order to slow the ripening of fruits. The first challenge was to control the levels of O_2 in the package. Since then, an enormous variety of polymers with different properties have been developed to offer a wide range of options in features such as gas permeability, tensile strength and flexibility, among others. Presently, diverse systems of modified atmosphere packaging have been developed and used with a wide range of fruits and vegetables in order to provide optimal storage conditions and product longevity. Table 1 summarizes the optimal conditions of temperature and gas composition of O_2 and CO_2 for the transport and / or storage of some fruits and vegetables.

2.1 Passive

Modified atmospheres can be obtained passively between plant material and sealed package or intentionally using determined concentrations of gases. Modified atmosphere is formed as a result of vegetable respiration, which consumes CO_2 and releases O_2 in sealed package. In passive modification, the respiring product is placed in a polymeric package and sealed hermetically. Only the respiration of the product and the gas permeability of the film influence the change in gaseous composition of the environment surrounding the product. If the product's respiration characteristics are properly matched to the film permeability values, then a beneficial modified atmosphere can be passively created within a package. The polymer itself variably restricts gas exchange between the internal and external environments due to its selective permeability to O_2 and CO_2. After a period of time, the system reaches an equilibrium atmosphere containing of lower concentrations of O_2 and higher concentrations of CO_2 than in atmospheric air.

Product	Temperature range (^{0}C)	Atmosphere	
		%O_2	%CO_2
Apples	0-5	1-2	0-3
Banana	12-16	2-5	2-5
Blackberry	0-5	5-10	15-20
Blueberry	0-5	2-5	12-20
Cherry, sweet	0-5	3-10	10-15
Cranberry	2-5	1-2	0-5
Grape	0-5	2-5	1-3
Kiwifruit	0-5	1-2	3-5
Lemon	10-15	5-10	0-10
Lychee (Lichti)	5-12	3-5	3-5
Mango	10-15	3-7	5-8
Nuts and dried fruits	0-10	0-1	0-100
Orange	5-10	5-10	0-5
Papaya	10-15	2-5	5-8
Persimmon	0-15	3-5	5-8
Pineapple	8-13	2-5	5-10
Plum	0-5	1-2	0-5
Raspberry	0-5	5-10	15-20
Strawberry	0-5	5-10	15-20
Artichoke	0-5	2-3	2-3
Asparagus	1-5	Air	10-14
Beans	5-10	2-3	4-7
Broccoli	0-5	1-2	5-10
Brussels sprouts	0-5	1-2	5-7
Cabbage	0-5	2-3	3-6
Cantaloupes	2-7	3-5	10-20
Cauliflower	0-5	2-3	3-4
Celery	0-5	1-4	3-5
Cucumbers	8-12	1-4	0
Herbs*	0-5	5-10	4-6
Lettuce	0-5	1-3	0
Onions	0-5	1-2	0-10
Parsley	0-5	8-10	8-10
Pepper	5-12	2-5	2-5
Radish	0-5	1-2	2-3
Spinach	0-5	7-10	5-10
Tomatoes	12-20	3-5	2-3

*Herbs: chervil, chives, coriander, dill, sorrel and watercress, Adapted from Kader (2002)

Table 1. Summary of optimal conditions for modified atmosphere and temperature during transport and / or storage of fruits and vegetables

2.2 Active

The concept of active packaging has been developed to adjust the deficiencies in passive packaging such as when a film is a good barrier to moisture, but not to oxygen, the film can still be used along with an oxygen scavenger to exclude oxygen from the pack. An intentionally or actively obtained modified atmosphere occurs when the desired gas mixture is introduced into the container before sealing. In this way, atmospheric balance inside the package is reached faster or almost immediately. Sometimes, certain additives are incorporated into the polymeric packaging film or within packaging containers to modify the headspace atmosphere and to extend shelf-life. Another process is the acceleration of atmospheric balance under partial vacuum packaging is the process of removing the air before sealing, reducing the free space. Although the active modification of the atmosphere within the package incurrs additional costs, the advantage is that the desired atmosphere is securely achieved in considerably less time.

2.3 O_2 and CO_2 limits

Safe levels of O_2 and CO_2 are important for package design. A lower O_2 limit has been associated with onset of fermentation and accumulation of ethanol and acetaldehyde (Beaudry et al., 1993). Fermentation is linked to the development of off-flavors and/or tissue damage. Effect of temperature on lower O_2 limit has been measured for a number of commodities including whole apple, apple slices, blueberry, and raspberry. In each case, lower O_2 limit increased with temperature. Lower O_2 limits vary from 0.15% to 5% and are influenced by temperature, commodity and cultivar (Beaudry and Gran, 1993).

It is necessary to know the main effects of gases on fresh fruits and vegetables and the interactions between gas and produce on the one hand and between gas and packaging material on the other hand to achieve the goal. It is important to recognize that while atmosphere modification can improve the storability of some fruits and vegetables, it also has the potential to induce undesirable effects. Fermentation and off-flavors may develop if decreased O_2 levels cannot sustain aerobic respiration (Kays, 1997). Similarly, injury will occur if CO_2 exceeds tolerable levels. Ranges of non-damaging O_2 and CO_2 levels have been published for a numbers of fruits and vegetables (Kader, 1997; Kupferman, 1997; Richardson and Kupferman, 1997; Saltveit, 1997; Beaudry 1999, 2000), minimally processed products (Gorny, 1997), and flowers and ornamentals (Reid, 1997). Horticultural crops differ in their tolerance for O_2 (Table 2) and CO_2 (Table 3).

3. Types of plastic films used in MAP

In a modified atmosphere packaging, changes start to take place immediately after packing the fresh produce as a result of the respiration of the packaged produce. The gases of the contained atmosphere and the external ambient atmosphere try to equilibrate by permeation through the package walls at a rate dependant upon the differential pressures between the gases of the headspace and those of the ambient atmosphere. It is in this context that the barrier to gases and water vapor provided by the packaging material must be considered. Thus, the success of the modified atmosphere packaging depends upon the barrier material used. MAP for fresh produce must also allow entry of oxygen to maintain the aerobic

O_2 (%)	Commodity
2	Lettuce (crisphead), pear
3	Artichoke, tomato
5	Apple (most cultivars), apricot, cauliflower, cucumber, grape, nashi, olive, orange, peach (clingstone), potato, pepper (bell)
7	Banana, bean (green snap), kiwi fruit
8	Papaya
10	Asparagus, brussels sprouts, cabbage, celery, lemon, mango, nectarine, peach, persimmon, pineapple, sweet corn
15	Avocado, broccoli, lychee, plum, pomegranate, sweetsop
20	Cantaloupe (muskmelon), durian, mushroom, rambutan
25	Blackberry, blueberry, fig, raspberry, strawberry

*Data are from Beaudry (2000), Gorny (1997), Kader (1997), Kupferman (1997), Richardson and Kupferman (1997), and Saltveit (1997)

Table 2. O_2 limits below which injury can occur for selected horticultural crops held at typical storage temperatures

CO_2 (kPa)	Commodity
< 0.5	Chopped greenleaf, Romaine and iceberg lettuce, spinach, sliced pear, broccoli
1.0	Broccoli florets, chopped butterhead lettuce, sliced apple, brussels sprouts, cantaloupe, cucumber, crisphead lettuce, onion, apricot, avocado, banana, cherimoya, sweet cherry, cranberry, grape, kiwifruit, litchi, peach, plum
2.0	Shredded and cut carrots, artichoke, cabbage, cauliflower, celery, bell and chili pepper, sweet corn, tomato, blackberry, fig, mango, olive, papaya, pineapple, pomegranate, raspberry, strawberry
2.5	Shredded cabbage, blueberry
3.0	Cubed or sliced cantaloupe, low permeability apples and pears, persimmon
4.0	Sliced mushrooms
5.0	Green snap beans, lemon, lime, orange
10.0	Asparagus
14.0	Orange sections

* Adapted from Herner (1987), Kader (1997), and Saltveit (1997)

Table 3. CO_2 partial pressures above which injury will occur for selected horticultural crops

metabolism of the product. In addition, some carbon dioxide must exit from the package to avoid build up of injurious levels of the gas. These packages are made of plastic films with relatively high gas permeability.

Packaging films that provide a wide range of physical properties, many of these individual films are combined through processes like lamination and co-extrusion. There are several groupings in MAP films such as in the plural, Vinyl Polymers, Styrene Polymers,

Polyamides, Polyesters and other polymers. Polypropylene is part of the Polyolefin group and used largely in MAP, in both forms: continuous and perforated. Sanz et al. (1999) studied the quality of strawberries packaged with polypropylene film, with proper perforations, during commercial postharvest practices. They concluded that perforated-mediated MA packaging helped to preserve fruit ripeness degree better, maintaining its nutritional value, measured as ascorbic acid content.

Many type of plastic films for packaging are available, but relatively few have been used to pack fresh fruits and vegetables, and even fewer have a gas permeability that makes them suitable for modified atmosphere packaging. The permeability of CO_2 should be 3 to 5 times the permeability of O_2. Many polymers used to formulate packaging films are within this criterion (Table 4).

Film type	Permeabilities (cc/m²/mil/dia a 1 atm)		
	CO_2	O_2	CO_2:O_2 ratio
Polyester	180-390	52-130	3.0-3.5
Polyethylene, low density	7,700-77,000	3,900-13,000	2.0-5.9
Polypropylene	7,700-21,000	1,300-6,400	3.3-5.9
Polystrene	10,000-26,000	2,600-7,700	3.4-3.8
Polyvinyl chloride	4,263-8,138	620-2,248	3.6-6.9

*Adapted from Kader (2002)

Table 4. Permeability for packaging fresh fruits and vegetables

4. Package parameters

Modified atmosphere (MA) packaging systems designed to produce optimum O_2 and CO_2 concentrations at suitable temperatures have been mathematically modeled (Chinnan, 1989; Lee et al., 1991; Exama et al., 1993; Talasila et al., 1995; Cameron et al., 1995).

The composition of the atmosphere inside the packaging results from the interaction of several factors including the permeability characteristics of the packaging material, the respiratory behavior of the plant material and the environment. Packaging films are selected so that the package has specific permeability characteristics and so that changes to these characteristics (temperature and humidity) over time, following the laws of physics. The concentration of gases within the packaging can be controlled to provide specific conditions. In contrast to these known and controllable factors are the often unknown and uncontrollable responses of the plant material. The plant species, the cultivar, cultural practices, the development stage, harvest management, the tissue type and post harvest handling all contribute and influence the response of the material to the atmosphere generated.

The scope of plant responses can be modified by the initial flow of gas before sealing the packaging as well as by the inclusion of chemical treatments to slow undesirable processes or decrease contamination. Each component of the packaging process needs to be examined separately to improve the understanding of what it contributes to potential packaging strategies.

Mathematical models can integrate the film permeability to O_2, CO_2 and H_2O, and the respiratory response of the commodity to O_2, in some cases, CO_2, along with its lower O_2 limit and upper CO_2 limit (Beaudry et al., 1992; Cameron et al., 1994; Lakakul et al., 1999; Fishman et al., 1996; Hertog et al., 1998). These models permit the identification of limiting features of the film, package design, and product and environment conditions.

The major factors to be taken into account while selecting the packaging materials are the type of package, the barrier properties needed (permeabilities of individual gases and gas ratios when more than one gas is used), physical properties of machinability and strength, integrity of closure (heat sealing), fogging of the film as a result of product respiration, printability and others.

4.1 Respiration rate

Respiration is a process in which chemical reactions oxidize lipids and carbohydrates to carbon dioxide and water to produce energy, while the organelle responsible for aerobic respiration known as mitochondria. Part of the released energy is stored as chemical energy adenosine triphosphate (ATP) and part is lost as heat. This complex process can be influenced by several intrinsic factors such as product size, variety, maturity, type of tissue and extrinsic factors such as temperature, concentration of O_2 and CO_2 and mechanical damage (Day, 1993).

Knowledge of the minimum required for O_2 aerobic respiration is very important to avoid that the anaerobic pathway is the predominant route of respiration, which causes the accelerated loss of product quality. In order to maintain food safety, it is important to know the potential hazards of each product, the permeability of the films and the rate of respiration of fruits and vegetables (Watada et al., 1996).

Since both the rate of respiration and the permeability of the film are sensitive to temperature variations, and respond to these changes differently, it is expected that the package under modified atmosphere remains determined only within an atmosphere given temperature range (Zagory, 2000).

The maximal rate of respiration for most fruit and vegetable products undergoes a 4- to 6-fold increase from 0 to 15 °C (Beaudry et al., 1992; Cameron et al., 1994, 1995; Lakakul et al., 1999). This means that product respiration increases at two or three times the rate of LDPE permeability, and thirty times the rate of perforation permeability with increasing temperature. When respiratory demand for O_2 increases faster than O_2 permeation as temperature increases, O_2 levels decline and may pose a risk to product quality. This limits the usefulness of MAP in some situations.

Variation in the respiration rate of the product and the variation in film or permeability can influence package design. Variation in product respiration and package permeability has been measured for broccoli and the effect on package O_2 modeled (Cameron et al., 1993). Cameron et al. (1993) concluded that there is an estimable risk of the package O_2 falling sufficiently low to promote fermentation in any product. Packages should be designed to generate O_2 levels well above the lower limit to ensure aerobic conditions. Products such as broccoli, mushrooms, leeks and others have very high rates of respiration, and most continuous films do not have the capacity to provide enough O_2 to avoid fermentation.

Accordingly, there is commercial interest to develop films with high gas transmission rates. Films that have improved rates of gas transmission by virtue of their polymeric nature are often blends of two or three different polymers, where each polymer performs a specific function such as strength, transparency and improved gas transmission. Similarly, films can be laminated to achieve needed properties.

4.2 Temperature

Temperature is one of the most important factors in extending the shelf-life of perishable products. Optimum storage temperature must be established for every product. Permeability of polymeric packaging films is also a function of temperature and it generally increases with the increase in temperature.

The effects of temperature on chemical reactions, including respiratory rate, traditionally quantified by Q_{10}, which is a coefficient by which it is possible to calculate how many times increases the rate of a reaction for each increase in temperature of 10 °C. The effect of temperature can also be quantified by the Arrhenius model, where the effect of temperature increase is given by the activation energy (Ea) (Cameron et al., 1995). The temperature quotient is useful because it allows us to calculate the respiration rates at one temperature from a known rate at another temperature. However, the respiration rate does not follow ideal behavior, and the Q_{10} can vary considerably with temperature. At higher temperatures, the Q_{10} is usually smaller than at lower temperatures.

The storage temperature is very important for the evolution of microbial and visual quality of fresh fruits and vegetables. Knowledge of weather conditions and temperature in the cold chain for fresh produce is needed to determine the influence of the cold chain in real loss of quality and shelf life of these products (Cortez et al., 2002).

The temperature of the produce in the package is managed by circulating cool air around the outside of the package. The film and the headspace atmosphere are barriers to heat movement, prevent rapid cooling, and reduce the effectiveness of refrigeration. A 'safe radius' for the distance from the center of the package to the circulated air can be calculated based on the heat of respiration and the rate at which heat can be removed by the cooler air (Sharp et al., 1993). For instance, the center of a package of broccoli must have a radius of less than 14 cm to keep it within 1 °C of the refrigerated air.

Temperature control, when combined with correct use of packaging and modified atmosphere technology, is effective in controlling metabolic processes described above. However, the ideal temperature handling, storage and marketing of fresh fruits and vegetable is generally not respected. Some examples of the ranges of temperature and relative humidity are to be respected in Table 1.

If it is necessary to choose between mild temperatures that cause symptoms of chilling injury and temperatures accelerate senescence and microbial growth, the first must be chosen (Watada and Qi, 1999).

4.3 Permeability

The permeability of the packaging material determines the atmospheric conditions in the headspace and ultimately the shelf-life of the product. If an atmosphere higher in carbon dioxide and / or lower in oxygen is required, the material should be impermeable to the gases. Vegetables and fruits require a certain amount of oxygen in the headspace for maintenance of quality, therefore, packaging material for these products should be quite permeable to the oxygen, to allow atmospheric oxygen to replenish the gas in the package.

Transparency of packaging material to light is also important.

The choice of packaging material is an extremely important part of the MAP operation. The materials must be cost effective, have low water vapor transmission rate, high gas barrier, mechanical strength to withstand machine handling and subsequent storage and distribution of the finished pack as well as have the capability of giving high integrity seals to ensure retention of gas within the pack until opened by consumer. Also, once a gas atmosphere is applied, the level and proportion of headspace gas/gases is controlled only by judicious selection of packaging material with specified permeability characteristics. . Thickness is also a factor controlling permeability.

There are two ways to create barriers using film. The first strategy employs continuous films that control the movement of gases in and out of the box. The second strategy uses film with small openings or microperforations.

With continuous films, the movement of O_2 and CO_2 is directly proportional to the differences in gas concentration across the film. Constant levels of gases in the package are achieved when the product's consumption of O_2 and production of CO_2 are equal. This situation exists only when the respiration rate is constant (Fishman et al., 1996).

In perforated films, the rate of movement of gas through the perforated film is the sum of gas diffusion and atmospheric air infiltration through the polymer film. Generally, total gas flow through the perforations is much greater than gas movement through the film. The rate of gas exchange through microperforations is much greater than through continuous films. Perforated packaging is more suitable for vegetables that have a high demand for O_2 (Gonzalez et al., 2008).

5. Respiration rate measurement

The respiration rate of fresh fruits and vegetables can be expressed as the rate of O_2 consumption and CO_2 production rate. The usual methods for determining the concentration of respiration are static or closed system, continuous flow system, and permeable system. The respiration rate is measured in permeable system, in other words, the product is within the dimensions and permeability of packaging film known (Beaudry, 1993; Joles et al., 1994; Lakakul, Beaudry, & Hernandez, 1999; Lee, Song, & Yam, 1996; Piergiovanni, Fava, & Ceriani, 1999; Smyth et al., 1998). The concentrations of O_2 and CO_2 are determined and stable mass balance is done on the system in order to estimate respiration rates.

The concentrations of gases in the system depend on the permeability characteristics of permeability, package size and weight of the product that were discussed previously. The time to reach equilibrium can be seen as a limitation of this method to measure the respiration rate. Definition of the steady-state concentration values is another difficulty of the permeable method.

For all methods have limitations, but this involves a greater number of variables, since the size of the packet are involved other parameters such as free volume, surface area and thickness of the gas exchange and the permeability characteristics. None of methods is clearly preferable over the others. When choosing the respiration rate determination method for a specific study, the benefits and limitations of each method should be taken into consideration. Fonseca et al. (1992) studied three different methods (closed, flow through and permeable systems) to measure respiration rate and observed their advantages and limitations. The main characteristics of these methods are summarized in Table 5.

Characteristics	System		
	Closed	Flow through	Permeable
Non-destructive	✓	✓	✓
Complexity of experimental set-up	✓	✓	✓
Ability to test different combinations of gases	Simple	Complex	Complex
Concentration is kept constant during the experiment	x	✓	✓
Suitable for low respiring products	✓	x	
Suitable for high respiring products	x	✓	✓
Accuracy is very sensitive to determination of	Free volume	Flow-rate	Permeability package dimensions, steady-state concentrations

*Adapted from Fonseca et al. (1992)

Table 5. Main characteristics of the three methods of respiration rate measurement

5.1 Factors affecting respiration rate

The factors affecting respiration are type and maturity stage of the commodity. Vegetables include a great diversity of plant organs such as roots, tubers, seeds, bulbs, fruits, sprouts, stems and leaves that have different metabolic activities and consequently different respiration rates. Even different varieties of the same product can exhibit different respiration rates (Fidler & North, 1967; Gran & Beaudry, 1992; Song et al., 1992). In general, non-climacteric commodities have higher respiration rates in the early stages of development that steadily decline during maturation (Lopez-Galvez, El-Bassuoni, Nie, & Cantwell, 2004). Respiration can also be affected by a wide range of environmental factors that include light, chemical stress, radiation stress, water stress, growth regulators, and pathogen attack. The most important postharvest factors are temperature, atmospheric composition, and physical stress.

One method to improve these problems would be to choose a film with permeability changes for O_2 similar to that of the respiration of the product, so if temperature increases, respiration and permeability of the film increase an equivalent amount.

5.2 Influence of temperature

Indeed the most important factor affecting postharvest life is temperature. This is because temperature has a profound effect on the rates of biological reactions, specifically metabolism and respiration. Over the physiological range of most crops, 0 to 30 °C increased temperatures cause an exponential rise in respiration.

Temperature of the product affects storability more than any other factor. Pre-cooling and temperature maintenance during handling and shipping are critical in preserving quality. Temperature also significantly affects film permeability and thereby the O_2 and CO_2 content of the package.

The elevated rate of respiration at high temperature could be used to rapidly establish the desired package atmosphere, but this would only be useful in the few situations in which it would be more important to rapidly establish the atmosphere than to slow physiological processes, eg., to reduce cut-surface browning.

Another solution to the MAP temperature problem is to develop a package system that senses either the environment or the physiological status of the enclosed product and responds by increasing the permeability to O_2 (Cameron et al., 1993). Such 'sense-and-respond' packaging is technically difficult to develop, and progress has only been conceptual at this time (Smyth et al., 1999). A third approach is to design packages to function at the highest temperatures typically encountered in the distribution and retail cool chain and, as far as possible, maintain control over the temperature of the packaged product, thereby adapting to the limitations imposed by the film. Most companies using MAP have adopted this simple solution. Generally, the lowest temperature feasible is maintained, since temperature has a much more significant influence on preserving quality than the application of low O_2 (Kays, 1997).

Reações metabólicas, entre elas a respiração, são reduzidas em 2 a 3 vezes para cada decréscimo de 10 °C na temperatura, o que permite retardar a maturação e a senescência do produto (Brecht, 1995).

5.3 Influence of gas composition

In MAP, the pack is flushed with a gas or a combination of gases. The common gases used are oxygen, nitrogen and carbon dioxide. Traces of carbon monoxide, nitrous oxide, ozone, argon, ethanol vapour and sulphur dioxide are also used. Minimum oxygen levels are used to pack plant under MA because oxygen can react with the fruits and vegetables resulting in the oxidative breakdown of them into their constitutive parts. Nitrogen is an inert gas. Carbon dioxide is responsible for the bacteriostatic and fungistatic effect in MA packaged fruits and vegetables. It retards the growth of moulds and aerobic bacteria. The inhibitory effect of carbon dioxide to micro-organisms is increased as the temperature is lowered.

The tolerance of any plant tissue at low O_2 tension is smaller as the storage temperature increases, since the requirements for aerobic respiration of the tissue increases with increasing temperature. Depending on the product damage associated with CO_2 can both increase and decrease with rise in temperature. CO_2 production increases with temperature, but its solubility decreases. Thus, the CO_2 concentration in the tissue may decrease or increase with increasing temperature. In addition, the physiological effect of CO_2 could be temperature dependent.

The activation energy is a parameter that has been used to characterize plastic packaging. Knowledge of the activation energy of the breath of product and packaging serves as an important tool to predict the effects of temperature fluctuations on the concentration of gases inside the package (Cameron et al., 1995).

6. Plant responses of MAP

Some of the most important factors that affect shelf life of fresh horticultural products are ripening and/or senescence, decay, and cut surface browning. The effect of modified atmosphere on these factors has been studied for different crops. The application of MA to affect these limiting factors can be restricted for some crops by adverse and/or non-beneficial physiological responses. Additionally, the good temperature management should be associated to by atmosphere modification.

Low O_2 and elevated CO_2 concentrations can significantly reduce the rates of ripening and senescence primarily by reducing the synthesis and perception of ethylene (Burg and Burg, 1967; Abeles et al., 1992). Changes in respiration and starch, sugars, chlorophyll, and cell wall constituents during this period can be reduced, and in some cases nearly arrested, by eliminating ethylene action through the use of low O_2/high CO_2 atmospheres.

Chlorophyll loss, a desirable trait for many climacteric fruits, results in quality loss for many vegetables. Chlorophyll degradation during the senescence of green vegetables can be reduced by low O_2 and elevated CO_2 (Ku and Wills, 1999).

Modified atmospheres are most effective at reducing ripening prior to the onset of ripening, rather than at a later stage. At the same time, packaged fruits and vegetables are usually intended for immediate consumption by the consumer and an unripe product is not immediately edible or is of reduced quality relative to the ripe product. Thus, the advantage of extend shelf-life by retarding ripening runs counter to the needs of the consumer when retail MAP systems are used. Nevertheless, MAP can reduce the rate of ripening of some commodities such as tomato even during its later stages (Yang and Chinnan, 1988).

Decay control is a particularly important problem for many crops. Levels of above 10% CO_2 effectively slow or stop the growth of numerous decay organisms (Brown, 1922). Low O_2 has a very limited effect on decay organism activity or survival at levels above the fermentation threshold of most commodities. Strawberry, blueberry, blackberry, raspberry and cherry, they are examples that can be stored at CO_2 atmosphere between 10 and 20%.

A negative plant responses to modified atmosphere packaging is when the respiration is reduced as O_2 becomes limiting. Although there is a limit to which O_2 can be reduced. The lower O_2 limit is frequently considered to be the level of O_2 that induces fermentation. However there are lower O_2 levels that may confer benefits that outweigh the loss in flavor or other quality parameters. Ethanol, acetaldehyde, ethyl acetate and lactate are products of fermentation that can contribute to the development of off-flavors (Kays, 1997; Mattheis and Fellman, 2000).

Synthesis of aroma compounds are generally suppressed by high CO_2 and low O_2 levels, in part by their action on ethylene sensitivity, but also via action of O_2 on oxidative processes, including respiration required for substrate production. But low O_2 MAP may suppress aroma production so consumers perceive reduced quality upon opening the container.

7. Conclusions

The benefits of MAP technology to the manufacturer, retailer as well as consumer far outweigh the drawbacks. Nevertheless some critical points should be considered in this technology. The following list some advantages and disadvantages of MAP.

- Advantages
  - Increased shelf-life allowing lesser frequency of loading of retail display shelves.
- Improved presentation of the product
  - Hygienic stackable pack sealed and free from product drip and odor
  - Shelf-life can be increase by 50 to 400%.
  - Reduction in production and storage costs due to better utilization of space and equipment.
- Disadvantages
  - Capital cost of gas packaging machinery
  - Increased pack volume increases transport costs and retail display space
  - Cost of gases and packaging materials
  - Temperature control is of critical importance and, by itself, has a greater impact than atmosphere modification for most products
  - Potential growth of food borne pathogens due to non-maintenance of required storage temperature by retailers and consumers.

8. References

Abeles, F. B., Morgan, P. W., & Saltveit, M. E. *Ethylene in plant biology*. 2nd ed. San Diego: Academic Press, 1992. 414 p.

Beaudry, R.M., & Gran, C. D. (1993). Using a modified-atmosphere packaging approach to answer some postharvest questions: Factors affecting the lower oxygen limit. *Acta Hort.*, Vol. 362, pp. 203-212.

Beaudry, R. M. (2000). Responses of horticultural commodities to low oxygen: limits to the expanded use of modified atmosphere packaging. *HortTechnology*, Vol. 10, pp. 491-500.

Beaudry, R. M. (1999). Effect of O_2 and CO_2 partial pressure on selected phenomena affecting fruit and vegetable quality. *Postharvest Biol. Technol.*, Vol 14, pp. 293-303.

Beaudry, R. M., Cameron, A. C., Shirazi, A, & Dostal-Lange, D. L. (1992). Modified-atmosphere packaging of blueberry fruit: Effect of temperature on package O_2 and CO_2. *Amer. Soc. Hort.* Vol. 117, pp. 436-441.

Beaudry, R. M., & Gran, C. D. (1993). Using a modified-atmosphere packaging approach to answer some postharvest questions: Factors affecting the lower oxygen limit. *Acta Hort.* Vol. 362, pp. 203-212.

Boyette, M.D., Sanders, D.C., & Rutledge, G.A. (1996). Package requirements for fresh fruits and vegetables. The North Carolina Agricultural Extension Service. North Carolina State University, Ralley, NC. USA. Publication no 9/96-3m-TWK-260373-AG-414-8.

Brecht, J. K. (1995). Physiology of lightly processed fruits and vegetables. *HortScience,* Vol. 30, No. 1, pp. 18-21.

Brown, W. (1922). On the germination and growth of fungi at various temperatures and in various concentrations of oxygen and carbon dioxide. *Ann. Bot.* Vol. 36, pp. 257-283.

Burg, S. P., & Burg, E. A. (1967). Molecular requirements for the biological activity of ethylene. *Plant Physiol,* Vol.42, pp. 114-152.

Cameron, A. C., Beaudry, R. M., Banks, N. H., & Yelanich, M. V. (1994). Modified atmosphere packaging of blueberry fruit: modeling respiration and package oxygen partial pressures as a function of temperature. *Journal of the American Society for Horticultural Science.* Vol. 119, no. 3, pp. 534-539.

Cameron, A. C., Patterson, B. D., Talasila, P. C., & Joles, D. W. (1993). Modeling the risk in modified atmosphere packaging: A case for sense-and-respond packaging. In: Proc. 6th Intl. Controlled Atmosphere Res. Conf., Ithaca, NY, pp. 95-112.

Cameron, A. C., Talasila, P. C., & Joles, D.J. (1995). Predicting the film permeability needs for modified atmosphere packaging of lightly processed fruits and vegetables. *HortScience* Vol. 30, pp. 25-34.

Cantwell, M. (2004).*Fresh-cut vegetables.* USA: University of California, Davis. p. 78-85. (Postharvest Horticulture Series n. 10.)

Chinnan, M. S. (1989). Modeling gaseous environment and physio-chemical changes in fresh fruits and vegetables in modified atmospheric storage. (In) "Quality factors of fruits and vegetables - Chemistry and Technology". J. J. Jen (ed.). ACS Symposium Series No. 405, p. 189-202. Washington, D.C.

Day, B. P. F. (1996). High oxygen modified atmosphere packaging for fresh prepared produce. *Postharvest News and Information,* Wallingford, v. 7, n. 3, p. 1N-34N.

Fidler, J. C., & North, C. J. (1967). The effect of conditions of storage on the respiration of apples. I. The effects of temperature and concentrations of carbon dioxide and oxygen on the production of carbon dioxide and uptake of oxygen. Journal of Horticultural Science, 42, 189-206.

Fishman, S., Rodov, V., & Ben-Yehoshua, S. (1996). Mathematical Model for Perforation Effect on Oxygen and Water Vapor Dynamics in Modified-Atmosphere Packages. *Journal of Food Science,* Vol. 61, No. 5, pp. 956–961.

Fonseca, S. C., Oliveira, F. A. R., & Brecht, J. K. (2002). Modelling respiration rate of fresh fruits and vegetables for modified atmosphere packages: a review. *Journal of Food Engineering,* v. 52, p. 99–119.

Fonseca, S.C., Oliveira, F.A.R., & Brecht, J.K. (2002). Modelling respiration rate of fresh fruits and vegetables for modified atmosphere packages: a review. *Journal of Food Engineering,* Vol. 52, pp. 99–119.

González, J., Ferrer, A., Oria, R., & Salvador, M. L. (2008). Determination of O_2 and CO_2 transmission rates through microperforated films for modified atmosphere packaging of fresh fruits and vegetables. *Journal of Food Engineering,* Vol. 86, No. 2, pp. 194-201.

Gorny, J. R. (1997). A summary of CA and MA requirements and recommendations for fresh-cut (minimally-processed) fruits and vegetables. In: J. Gorny (ed) Fresh-cut fruits and vegetables and MAP. Postharvest Hort. Series No. 19, Univ. Calif., Davis CA, CA'97 Proc. 5.30-66.

Gorris, L., & Tauscher, B. (1999). Quality and safety aspects of novel minimal processing technology. *Processing of foods: Quality optimization and process assessment.* CRC Press, USA, pp. 325-339.

Herner, R.C. (1987). High CO_2 effects on plant organs. In: J. Weichman (ed) Postharvest Physiology of Vegetables, Marcel Dekker, NY, pp. 239-253.

Hertog, M.L.A.T.M., Peppelenbos, H. W., Evelo, R.G., & Tijskens, L.M.M. (1998). A dynamic and generic model of gas exchange of respiring produce: the effects of oxygen, carbon dioxide and temperature. *Postharvest Biol. Technol.* Vol. 14, PP. 335-349.

Joles, D. W., Cameron, A. C., Shirazi, A., Petracek, P. D., & Beaudry, R. M. (1994). Modified atmosphere packaging oh 'Heritage' red raspberry fruit: respiratory response to reduce oxygen, enhanced carbon dioxide and temperature. *Journal of the American Society for horticultural Science.* Vol. 119, No. 3, pp. 540-545.

Kader, A.A. (2002). *Post-harvest technology of horticultural crops.* Oakland: University of California, Division of Agriculture and Natural Resources Publication 3311, 535 pp.

Kader, A.A. (1997). A summary of CA requirements and recommendations for fruits other than apples and pears. In: A. Kader (ed) Fruits other than apples and pears. Postharvest *Hort. Series* No. 17, Univ. Calif., Davis CA, CA'97 Proc. 2:1-36.

Kader, A. A., Zagory, D., & Kerbel, E. L. (1989).Modified atmosphere packaging of fruits and vegetables. *Rev. Food Science and Nutrition,* Vol. 28, No. 1, pp. 1-30.

Kays, S. J. (1997). Postharvest physiology of perishable plant products. Van Nostrand Reinhold, NY.

Ku, V. V. V., & Wills, R. B. H. (1999). Effect of 1-methylcyclopropene on the storage life of broccoli. *Postharvest Biol. Technol,* Vol. 17, pp. 127-132.

Kupferman, E. (1997). Controlled atmosphere storage of apples. In: E.J. Mitcham (ed) Apples and Pears. *Postharvest Hort. Series* No. 16, Univ. Calif., Davis CA, CA'97 Proc. 3:1-30.

Lakakul, R., Beaudry, R. M., & Hernandez, R. J. (1999). Modeling respiration of apple slices in modified-atmosphere packages. *Journal of Food Science.* Vol. 64, No. 1, pp. 105-110.

Lee, D. S., Haggar, P. E., Lee, J., & Yam, K. L. (1991). Model for fresh produce respiration in modified atmospheres based on principles of enzyme kinetics. *Journal of food Science,* Vol. 56, No. 6, pp. 1580-1585.

Mattheis, J.P., Fellman, & J.P. (2000). Impact of modified atmosphere packaging and controlled atmosphere on aroma, flavor and quality of horticultural produce. *HortTechnology,* Vol. 10, pp. 507-510.

Mir, N., & Beaudry, R. M. (1986). Modified Atmosphere Packaging. USDA Handbook 66. Washington, D.C. GPO. 7/11/2011, Available from: <http://www.ba.ars.usda.gov/hb66/015map.pdf>.

Reid, M.S. (1997). A summary of CA and MA requirements and recommendations for ornamentals and cut flowers. In: M.E. Saltveit (ed) Vegetables and ornamentals. *Postharvest Hort. Series* No. 18, Univ. Calif., Davis CA, CA'97 Proc. 4:129-136.

Richardson, D. G., & Kupferman, E. (1997). Controlled atmosphere storage of pears. In: E.J. Mitcham (ed) Apples and pears. *Postharvest Hort. Series* No. 16, Univ. Calif., Davis CA, CA'97 Proc. 2:31-35.

Saltveit, M. E. (1997). A summary of CA and MA recommendations for harvested vegetables. In: M.E. Saltveit (ed) Vegetables and ornamentals. *Postharvest Hort. Series* No. 18, Univ. Calif., Davis CA, CA'97 Proc. 4:98-117.

Sanz, C., Pérez, A.G., Olías, R., & Olías, J.M. (1999). Quality of Strawberries Packed with Perforated Polypropylene. *Journal of Food Science*, Vol. 64, No. 4, pp. 748-752.

Sharp, A. K., Irving, A.R., & Morris, S.C. (1993). Does temperature variation limit the use of MA packaging during shipment in freight containers? In: G. Blanpied, J. Bartsch and J. Hicks (eds) Proc. 6th Intl Contr. Atmos. Res. Conf., Cornell Univ., Ithaca NY, pp. 238-251.

Shirazi, A., & Cameron, A.C. (1992). Controlling relative humidity in modified-atmosphere packages of tomato fruit. *HortScience*, Vol. 27, pp. 336-339.

Smyth, A. B., Song, J., & Cameron, A. C. (1998). Modified atmosphere packaged cut iceberg lettuce: effect of temperature and O_2 partial pressure on respiration and quality. *Journal of Agricultural and Food Chemistry*, Vol. 46, pp. 4556-4562

Smyth, A. B., Talasila, P. C., & Cameron, A. C. (1999). An ethanol biosensor can detect low-oxygen injury in modified atmosphere packages of fresh-cut produce. *Postharvest Biol. Technol*, Vol. 15, pp.127-134.

Talasila, P. C., Chau, K. V., & Brecht, J. K. (1995). Modified atmosphere packaging under varying surrounding temperature. *Trans. ASAE*,Vol. 38, pp. 869–876.

Tano, K., Oulé, M. K., Doyon, G., Lencki, R. W., & Arul, J. (2007). *Postharvest Biology and Technology*, Vol. 46, pp. 212–221.

Yang, C. C., & Chinnan, M.S. (1988). Modeling the effect of O_2 and CO_2 on respiration and quality of stored tomatoes. *Trans. ASAE*, Vol. 31, pp. 920-925.

8

Study of Adhesion and Surface Properties of Modified Polypropylene

Igor Novák[1], Anton Popelka[1], Ivan Chodák[1] and Ján Sedliačik[2]
[1]Polymer Institute, Slovak Academy of Science,
[2]Technical University in Zvolen,
Slovakia

1. Introduction

Isotactic polypropylene (iPP) is one of the most frequently applied polymers. Unfortunately, it has a disadvantage of being non-polar which makes the adhesive joints formed with other, more polar polymers, rather week (Brewis & Mathieson, 2002; Chodák & Novák, 1999; Pocius, 1997; Kinloch, 1987; Schultz & Nardin, 1994; Kolluri, 1994; Denes & Manolache, 2004). From the point of view of application of iPP it is desirable that their adhesive properties are improved (Yalizis et al.,2000; Ohare et al., 2002; Shenton et al., 2001; Kim et al., 2002; Moosheimer & Bichler, 1999). This demand can be met by a modification of iPP, while polymer can be modified either at the surface directly, e.g. by plasma of electric discharge (Denes & Manolache, 2004), or in bulk, by addition of a suitable polar low- or high-molecular compound to the polymeric matrix (Novák & Florián, 1994; Novák et al., 2004; Novák & Florián, 2001; Novák, 1996). The presented contribution aims at offering the efficient methods of surface modification of iPP, from the point of view of improving its adhesion properties. Recent achievements will be discussed to evaluate physical and chemical changes taking place as a result of modification.

The analysis of the low adhesive properties of iPP leads to the two different approaches of explanation (Brewis & Mathieson, 2002; Chodák & Novák, 1999, Kinloch, 1987). By the first explanation the low adhesion of iPP consists in a formation of thin layer of low-molecular substances on the interfacial boundary. The primary function of modification is then a removal of the thin low-molecular substance layer from the polymer surface, while the chemical modification itself is of a secondary importance. The second explanation attributes the low adhesive properties of iPP to its non-polar character and low surface energy, stressing the dependence of the adhesive properties of iPP on their super molecular structure. The chemical changes resulted in the increase of the polarity and surface energy are considering for the most important in the modification of iPP.

Low-molecular substances present in iPP (Brewis & Mathieson, 20021; Kinloch, 1987) such as antioxidants, lubricants, impurities introduced into polymer during polymerization process, the ends of polymer chains originating from the initiator, as well as impurities introduced into polymer during granulation and foil processing reduce the strength of adhesive joints due to their low cohesion. After removing the low- molecular substances by precipitation, the resulting polymer dispatched the higher adhesive properties. For iPP with wide distribution of molecular masses polymer-homologues with lower molar mass and higher mobility penetrate to the region of interfacial boundary by diffusion, reducing thus the interfacial tension. As a result of the thermal and Brown's motions, the interfacial boundary is enriched by oligomers and the value of the surface energy is reduced. In the region of iPP surface after cross linking in inert atmosphere by UV irradiation short polymer chains became cross linked removing thus the thin adhesion reducing layer of low-molecular substances on the interfacial boundary and as a result of this modification an increase of adhesive properties was observed.

IPP as a non-polar polymer has a very low (near zero) value of the polar component (PC) of the surface energy (Brewis & Mathieson, 2002; Chodák & Novák, 1999). When the polarity of iPP is increased also the value of the surface energy and its PC are increased as well. The growth of iPP polarity can be reached by utilization of some of existing methods of modification (Kinloch 1987), e.g. by flame modification. In the course of modification of iPP suitable polar functional groups are introduced. This process results into an increase of the compatibility of iPP with the polar polymers and the growth of its adhesion. Increasing the polarity of iPP can also be obtained by the addition of polar low-molecular substances or polymer to the matrix of iPP. In the case of this modification the increase of adhesion is to a large extent caused by an increasing mobility of polymer chains, e.g. by addition of fatty acids and their salts or oxidized and/or maleic anhydride grafted paraffin wax to iPP. From the point of view of the diffusion theory of adhesion (Schultz & Nardin, 1994) it is the kinetic effect taking part in this increase.

The iPP has in comparison with low-density polyethylene a higher degree of crystallinity (Brewis & Mathieson, 2002; Kinloch, 1987). During crystallization of iPP a smectic or a crystalline monoclinic form is formed depending on the thermal history of the sample. The smectic form is relatively stable up to temperature 50 ᵒC; at higher temperature it is transformed into the monoclinic crystalline form. It was experimentally proved that the adhesive properties of monoclinic crystalline form are higher than those of smectic form, while the value of the surface energy of iPP grows with the degree of changes from smectic form to monoclinic crystalline form.

The presence of low-molecular substances in iPP results in a formation of thin layer on the interfacial boundary decreasing the value of the surface energy. This effect prevails over the effect of increasing surface energy during the change of smectic form to crystalline monoclinic form and thus the adhesion of iPP is decreased after conditioning of adhesive joints at the temperature of 50 - 90 ᵒC. The increased adhesion of iPP can be reached by heating the polymer above the temperature of 170 ᵒC (or above the melting point of iPP) and subsequent fast cooling. Utilization of this procedure inhibits formation of the thin interfacial layer of low-molecular substances and iPP crystallized in the smectic form.

During the surface modification of iPP the structure of its surface changes and becomes rough. The changes of the surface structure during the modification of iPP depend on the applied modification method and on the thermal history of the polymer sample. It was found that there is a linear relationship between the roughness of the polymeric surface and the value of the mechanical work of adhesion. An increase of the mechanical properties of iPP is in this case related to mechanical fixation of adhesive in pores, or to micro defects of the polymeric surface. The change of the physical character of the surface takes place also in modification of iPP in mass by addition of suitable additives increasing iPP adhesion accompanied by simultaneous changes of super molecular structure of polymer.

IPP relatively easily undergoes the oxidation (Chodák & Novák, 1999). This is observed under the effect of light, heat, UV irradiation, plasma of electric discharge and oxidation agents while functional groups are formed which are capable of participating in the further chemical reactions (Brewis & Mathieson, 2002; Kinloch, 1987; Kolluri, 1994). In the course of oxidation of iPP hydroperoxide, carbonyl, carboxyl and hydroxyl groups are formed. In previous studies a formation of ozonides and peroxides during oxidation has been observed. In the case of halogenization methods of iPP modification halogens are attached to the polymer chain, and are often combined with the oxygen.

2. Experimental

2.1 Used polymers

For corona discharge modification experiments these sorts of iPP have been used:

- biaxially oriented foils of iPP Mosten 59 4928 (Chemopetrol, Czech Republic): thickness = 0.02 mm, density (23 °C) = 0. 956 $g.cm^{-3}$, T_{melt} = 170 °C, ΔH_{melt} = 94.4 $J.g^{-1}$. Extruded iPP of 0.02 mm thickness (Chemopetrol, Czech Republic): density (23 °C) = 0.905 $g.cm^{-3}$, T_{melt} = 165 °C, ΔH_{melt} = 49.4 $J.g^{-1}$. iPP backing fabric produced from Mosten 58 512 (Chemopetrol, Czech Republic) containing 0,2 wt.% of 2,6 - di (tertbutyl) -1,4 - dimethyl phenol: density (23 °C) = 0.920 $g.cm^{-3}$, T_{melt} = 162 °C, ΔH_{melt} = 89.7 $J.g^{-1}$. Aqueous dispersion of butyl acrylate-vinyl acetate copolymer Duvilax KA-31 (Duslo, Slovakia) containing 53.1 wt.% of dry content, pH = 4 - 6, content of free monomer 0.9 wt.%, viscosity 10 - 70 $mPa.s^{-1}$, was used as a polymer deposit on iPP backing fabric.

For grafting experiments were used these polymers as following:

- stabilised iPP Mosten 58 512 (Chemopetrol, Czech Republic) containing 0,2 wt.% of 2,6-ditertbutyl-1,4-dimethylphenol: density (23 °C) = 0.920 $g.cm^{-3}$, T_{melt} = 162 °C, ΔH_{melt} = 89.7 $J.g^{-1}$, polyamide Silamid 30 SW13 (PCHZ, Slovakia), polyvinyl acetate (Polysciences, USA) and cellulose acetate (Czech Republic), aluminum (Slovakia).

For modification by chromyl chloride and chromo sulfuric acid a stabilized polymer iPP Tatren TF- 411, M_v = $2.16.10^5$ (Slovnaft-Mol, Slovakia) was used.

The halogenization experiments were performed using a commercial iPP:

- iPP Tatren FD 420 containing 0.2 wt.% of UV stabilizer (Slovnaft-Mol, Slovak Republic), MFI = 3.2 g/10min, density (23 °C) = 0.905 $g.cm^{-3}$, polyvinyl acetate, M_w = 1.6 × 10^5 $g.mol^{-1}$ (Polysciences, USA).

2.2 Measurements methods

2.2.1 Contact angles and surface energy

The values of contact angles were determined by direct goniometric measurement using a Contact Angle Meter Amplival Pol (Zeiss, Germany). The drops of testing liquid (V = 3 µl) were introduced onto the polymeric surface. Each measurement was carried out with a set of the testing liquids: glycerol, formamide (Serva, Germany), thio diglycol, ethylene glycol, α-bromo naphthalene (Aldrich, USA), methylene iodide (Fluka, Switzerland), benzyl alcohol, diethylester of anthranil acid (Merck, Germany), twice distilled water. Each measurement was repeated 5 times at 25 ºC and the dependencies θ = f (t) were extrapolated to t = 0. The dispersive and PC of the surface energy of polymer were calculated according to the relation (47-50):

$$\frac{(1+\cos\theta)\cdot(\gamma_{LV}^d+\gamma_{LV}^p)}{2}=(\gamma_{LV}^d\cdot\gamma_s^d)^{1/2}+(\gamma_{LV}^p\cdot\gamma_s^p)^{1/2} \tag{1}$$

where θ is the contact angle (deg), γ_{LV}^d, γ_{LV}^p is the polar and dispersive component of the surface energy of testing liquid (mJ.m^{-2}), while

$$\gamma_s^{total}=\gamma_s^p+\gamma_s^d \tag{2}$$

where γ_s^{total} is the total surface energy and γ_s^p, γ_s^d is the polar and dispersive component of the surface energy of polymer (mJ.m^{-2}).

2.2.2 Interfacial equilibrium work of adhesion and interfacial tension

Interfacial equilibrium work of adhesion was calculated using the values of polar and dispersive components of the surface energy according to the relation:

$$W_{a1,2}=\frac{4\cdot(\gamma_1^d+\gamma_2^d)}{\gamma_1^d+\gamma_2^d}+\frac{4\cdot(\gamma_1^p+\gamma_2^p)}{\gamma_1^p+\gamma_2^p}=W_{a1,2}^d+W_{a1,2}^p \tag{3}$$

$$\gamma_{1,2}=\gamma_1+\gamma_2-W_{a1,2} \tag{4}$$

where $W_{a1,2}^d$ and $W_{a1,2}^p$ are the dispersive and polar component (PC) of interfacial equilibrium work of adhesion, (mJ.m^{-2}), γ_1^d, γ_2^d are the dispersive components of the surface energy of the polymer 1 and 2 respectively (mJ.m^{-2}), γ_1^p, γ_2^p are the PC of the surface energy of the polymer 1 and 2 respectively (mJ.m^{-2}), $\gamma_{1,2}$ is the interfacial tension between polymer 1 and polymer 2 (mJ.m^{-2}).

2.2.3 Mechanical work of adhesion by peeling

The mechanical work of adhesion was measured by peeling of adhesive joints using an 5 kN universal testing machine Instron 4301 with a peeling wheel with adjustable angle of peeling 90º at speed 10 mm.min^{-1}. The adhesive joints have been prepared by the application of polyvinyl acetate from ethyl acetate solution onto the technical cotton fabric "Molino" using a coating pad Dioptra (Czech Republic). The iPP foil and impregnated cotton fabric have been fixed together and the adhesive joints have been dried at 60 ºC to a constant weight, then they were cut to strips with dimensions 25 x 200 mm. The peeling length of an adhesive joint was 100 mm. The values of mechanical work of adhesion A_m (J.m^{-2}) were calculated by equation:

$$A_m=F_s/b \tag{5}$$

where F_s is the medium peeling force (N) and b is the width of the adhesive joint (m).

Measurements were carried out using a software developed for universal testing device Instron, Series IX.

2.2.4 Strength of adhesive joint by shear

The strength of adhesive joint was determined by testing of single overlapped adhesive joints in shear using a dynamometer Instron 4301 with aluminum slabs having dimensions 10 x 60 mm, the thickness was 2 mm, the length of overlapping area was 15 mm and the thickness of deposited adhesive was 0.1 mm.

2.2.5 Molecular weight

The molecular weight of iPP was determined on a high-temperature viscometer heated with silicone oil. The viscosity of iPP was measured in α-chloro naphthalene at 145 °C. The molecular weight was calculated from the following equation:

$$[\eta]=4.9 \cdot 10^{-3} \cdot M_v^{0.80} \qquad (6)$$

where $[\eta]$ is limiting viscosity number (dl.g^{-1}) and M_v is molecular weight of polymer obtained viscometrically.

2.2.6 ESR analysis

IPP modified by chromyl chloride was subjected to electromagnetic spin resonance measurements using a spectrophotometer E4. The following parameters were applied: frequency 100 kHz, microwave power 200 mW, resolution 100 mG, sensitivity 5 x 10^{10} spin.G^{-1}. The relative concentration of Cr^{IV} was calculated according to the equation:

$$C_{rel} = \frac{I \cdot \Delta H^2}{m \cdot s} \qquad (7)$$

where I, H, m and s are intensity of spectrum, width of band, amount of weighed sample and sensitivity of device respectively.

2.2.7 Hydroperoxides determination

Hydroperoxides content was determined spectrophotochemically. A sample of modified iPP was kept for 12 hours in vacuum for removing the physically adsorbed oxygen. Acetic acid and n-heptane were distilled and nitrogen was allowed to bubble through solutions. Then 3 ml of saturated solution of potassium iodide in isopropyl alcohol, 1 ml of glacial acetic acid and 1 ml of n-heptane were placed into a 10 ml Erlemayer flask.

The samples of foils of the modified iPP (2 x 2 cm) were put into the solution and left for 24 hours in inert atmosphere of carbon dioxide (dry ice). The absorbance at λ = 365 nm was measured with respect to blank experiment by using a UV spectrometer. The concentration of hydroperoxides was calculated from the expression:

$$A = \varepsilon \cdot c \cdot d A \qquad (8)$$

where A, ε, c and d are the absorbance, molar extinction coefficient (ε = 25 x 10^3 kg.mol^{-1}.cm^{-1}), concentration of hydroperoxides (mol O$_2$.kg^{-1} polymer) and thickness of absorption cell (cm) respectively.

2.2.8 Carbonyl groups determination

The method of determination of carbonyl groups in oxidized iPP was based on UV spectrophotometric determination of 2,4-dinitrophenylhydrazones by means of the absorption band at λ = 365 nm. Hydrazones originated in the reaction of oxidized iPP with 2,4-dinitrophenylhydrazine.

The reagent solution included 1 g of 2,4-dinitrophenylhydrazine, 100 ml of ethanol, 5 ml of hydrochloric acid and 5 ml of distilled water. Before each measurement a fresh solution of 2,4-dinitrophenylhydrazine was prepared. The formation of hydrazones proceeded at 70 °C in the course of 20 minutes. Then iPP was repeatedly washed with hot ethanol and perfectly dried. The disappearance of the infrared absorption bands corresponding to carbonyl groups supports the assumption that these groups were quantitatively consumed in the reaction with 2,4-dinitrophenylhydrazine.

The reaction with 2,4-dinitrophenylhydrazine finished the modified iPP was used for molding the foils that were liable to UV spectrophotometric measurements. Di (heptadecyl) ketone was used as model substance. It was added in the amount of 1wt.% to iPP, exposed to the reaction with 2,4-dinitrophenylhydrazine and thus a sample of standard was formed. According to the Beer's law it is valid:

$$\varepsilon = \frac{A_1}{c_1 \cdot d} \tag{9}$$

where ε, A_1, c_1 are stand for extinction coefficient, absorbance, the known concentration of carbonyl groups in sample of standard and width of foil, respectively.

The value of extinction coefficient calculated from equation (9) was used for calculation of the concentrations of carbonyl groups:

$$c_x = \frac{A_x}{\varepsilon \cdot d} \tag{10}$$

where A_x, c_x and d are absorbance of the investigated sample, the determined concentration of carbonyl groups and the width of the foil, respectively.

2.3 Modification methods

2.3.1 Modification by corona discharge

The modification of polypropylene foils by corona discharge was performed in a pilot plant (Softal 2005, Germany) in the medium of air oxygen at atmospheric pressure and temperature of 295 K. The used cylindrical electrodes were of 98 mm. The electrode voltage was equal to 9000 V and current density varied in the range among 0.3 and 0.8 mA.

2.3.2 Grafting by itaconic acid (IA)

Grafting of onto iPP was performed in an extruder in the polymeric melt at 220 °C during 180 sec. Itaconic acid (IA) concentration was up to 5 wt.%. Benzoyl peroxide (Aldrich, England) was added as an initiator and its concentration varied between 0.1 and 1 wt.%. Unreacted IA was then removed by Soxhlet extraction in boiling n-heptane. Pellets of IA modified iPP have been obtained. The pellets were compression molded at 210 ºC to films 0.05 mm thick suitable for goniometric measurements of contact angles.

2.3.3 Oxidation by chromyl chloride and chromo sulfuric acid

The modifications of iPP by chromyl chloride was accomplished in chromyl chloride vapor. The samples of iPP were individually prepared in Petri dishes and processed at individual temperatures by using particular reaction times. The reactions of oxidized iPP with 2,4-dinitrophenylhydrazine took place in a solution containing 1 gram of 2,4-dinitrophenylhydrazine as well as hydrochloric acid, ethanol and water in volume ratio 5 : 100 : 5.

Powdered iPP was modified by chromo sulfuric acid comprising sulfuric acid 96 %, potassium dichromate and water in weight ratio 100 : 5 : 8. Different reaction times and different temperatures were used for this method of modification.

2.3.4 Halogenization by UV/ phosphoryl chloride

The LDPE foils were placed in the glass vessel in the atmosphere of saturated vapors of phosphoryl chloride. The sorption of the UV sensitizer onto the surface of the PO foils took 24 hours. The foils were then irradiated by UV radiation in an open quartz tube having a diameter of 50 mm. The UV light with a wavelength λ = 366 nm was emitted by a 400 W mercury discharge lamp. The distance (d) of the UV tube from the surface of polymer varied between 50 and 200 mm. The relative intensity of UV radiation is inversely proportional to the square ratio of the UV source distances according to the relation:

$$I_r = \frac{I}{I_0} = (d_0/d)^2 \tag{11}$$

where I_r is the relative intensity of UV radiation, I_0 (W.s/m²) is the intensity of UV radiation at the reference distance d_0 = 50 mm and I (W.s/m²) is the intensity of UV radiation at the distance d (mm).

3. Results and discussion

3.1 Modification by corona discharge

During the modification of iPP by electric discharge plasma the surface of polymer interacts with the particles of plasma (Yalizis et al., 2000; OHare et al., 2002; Shenton et al., 2001; Kim et al., 2002; Novák et al., 2004; Noeske et al., 2004; Guimond et al., 2002; Mikula et al., 2003; Strobel et al., 2003; Poncin-Epaillard, 2002; Štefečka et al., 2003; Ráhel et al., 2003; Lehocký et al., 2003; Drnovská et al., 2003; Kuzuya et al., 2003). The active particles of plasma can originate either from the electric discharge at the atmospheric pressure (corona discharge) in air or from high (radio) frequency (RF) discharge at lower pressure in an inert atmosphere, or other atmosphere. The active particles of discharge plasma transferring the energy during the modification are ionic and neutral particles, electrons and UV radiation. Energy captured from the particles by the surface initiates chemical reactions leading to changed polarity and adhesion properties of iPP. The effect of plasma on the polymer surface includes the act of photons of UV irradiation having important role during the formation of radicals. Electrons and ions do not participate in the surface modification directly but are crucial for the formation of excited atoms and molecules in the volume of the discharge and these particles in contacts with the surface induce a formation of free radicals and chemical reactions. In the interaction of polymer surface with discharge plasma of inert gas, O, N, or the air the following reactions are effective (Kinloch, 1987; Kolluri, 1994):

a) direct reactions of active particles of plasma (surface oxidation by oxygen atoms in oxygen plasma or in the air), b) creating of free radicals and their consequent reactions (polymer degradation, grafting copolymerisation, incorporation of oxygen). The oxidation effect of corona discharge on the surface of the polymer can be schematically represented by following equations:

$$RH + 2O^{\bullet} \rightarrow R^{\bullet} + H^{\bullet} + O_2 \tag{12}$$

$$RH \xrightarrow[UV]{hv} R^{\bullet} + H^{\bullet} \tag{13}$$

Equation (12) represents a dissociation enabled by the absorption of dissociation energy of oxygen and equation (13) of dissociation by absorption of UV energy.

During the modification of iPP by the discharge plasma following reactions take place:

$$R^{\bullet} + O_2 \rightarrow ROO^{\bullet} \tag{14}$$

$$ROO^{\bullet} + RH \rightarrow ROOH + R^{\bullet} \tag{15}$$

$$ROOH \rightarrow RO^{\bullet} + OH^{\bullet} \tag{16}$$

Equations (14) and (15) represent the formation of hydroperoxides during the modification, equation (16) on the other hand represents the decomposition of hydroperoxides and the formation of carbonyl groups. For the modification of iPP the modification by corona discharge at atmospheric pressure as well as a high-frequency discharge with frequency 13.56 MHz at low pressures (RF discharge) is used most often. The following scheme to represent the effect of corona discharge on iPP has been proposed:

(17)

whereupon between enolic hydrogen of enolyzed ketogroup and other ketogroup in the macromolecule hydrogen bonds are formed.

$$
CH_3 \quad\underline{\quad\quad}\quad \overset{\displaystyle \sim CH}{\underset{\displaystyle \|}{C}} \quad\underline{\quad\quad}\quad OH \cdots\cdots C \overset{\displaystyle CH_2 \sim}{\underset{\displaystyle CH_3}{\diagdown}} \diagup \tag{18}
$$

In the case of humid environment during the corona treatment of iPP the reaction of radical with oxygen is competing with a reaction of radical with water according to the scheme:

$$
CH_2 \underline{\quad} \overset{\displaystyle \cdot}{\underset{\displaystyle |}{C}} \underline{\quad} CH_2 \xrightarrow[\ OH\ ]{(\ H_2 O)} \sim CH_2 \underline{\quad} \overset{\displaystyle OH}{\underset{\displaystyle |}{C}} \underline{\quad} CH_2 \sim \tag{19}
$$

$$
\underset{\displaystyle CH_3}{} \qquad\qquad\qquad\qquad \underset{\displaystyle CH_3}{}
$$

at which compounds are formed which do not contribute to the adhesive joint. A stronger decay was observed for the iPP polymerized in the discharge plasma than for that prepared by a conventional method.

The adhesion properties of iPP modified by high-frequency electric discharge in the air, oxygen and in nitrogen have been studied (Strobel et al., 2003). It was observed that after 30 seconds of exposure of the polymer in oxygen plasma the concentration of hydroperoxides reached 1.6×10^{-5} mol O_2/kg^{-1} iPP and the contact angle of redistilled water decreased from 94° to 40°. By further treatment by plasma the changes in contact angles periodically oscillated around the value 52°, while by the aging of treated samples the contact angle was increasing.

An important group of electric discharge methods is the modification by corona discharge (Yalizis et al., 2000). The treatment of iPP by corona discharge is strongly affected by humidity. By identification of chemical changes in iPP modified in dry air via ATR spectroscopy the presence of following functional groups was shown: C=O, OH, C=C, and COOH, and observed >C=O group to reach the highest concentration. The PC of the surface energy of iPP modified by corona discharge increased non-linearly with the degree of modification while the dispersion component is practically not changed. A decrease of temperature during the modification leads to a decrease of PC of the surface energy.

The adhesive properties of iPP modified by corona discharge are decreasing if a polymer surface makes contact with a compound forming the hydrogen bonds (most often a liquid). The formation of adhesive forces on the modified surface of iPP is related to the hydrogen bond of enolic group on one polymer surface and a similar polar group (e.g. carbonyl) on the other polymer surface and by heating up the surface or by the effect of heat formed by friction one can cancel the effect of modification and original good adhesive properties decrease.

The correlation between the adhesion of corona discharge modified iPP and the formation of hydroperoxides has also been studied (Novák & Florián, 2001; Novák & Florián, 2004). A linear increase of concentration of hydroperoxides and the PC of the surface energy with the current density of corona discharge has been found. At the same time it was shown that the efficiency of the surface treatment was strongly influenced by the super molecular structure of iPP: for extruded iPP foil a higher adhesion was found than for a biaxially oriented one. Furthermore by the study of a time elapsed from the modification they found that the adhesion of extruded polymer decreases with the time while the adhesion of biaxially oriented iPP does not vary with the time. The detail analysis of chemical changes of iPP modified by corona discharge is given also by other authors whose point to the fact that the reactive groups formed at the surface modification can participate in the further reactions.

A significant increase of adhesion of iPP was obtained by grafting a suitable monomer such as acryl amide (Novák & Florián, 1995) on the surface of iPP modified by discharge plasma. A very effective is the grafting mechanism with free radicals of the polymer in the case when the free radicals are formed relatively slowly. For the surface energy of biaxially oriented iPP modified by corona discharge and grafted by acryl amide a linear dependence on the current density was found (Lei et al., 2000). At the same time, after the grafting a decrease of concentration of hydroperoxides exhibited the result of linking of acryl amide on the active centers of modified iPP. In the case of modification of PO a time factor negatively affects the adhesion properties. The efficiency of modification decreases with the time elapsed from modification. For the given reasons a condition for the formation of strong adhesive bonds is an immediate processing of iPP after the modification.

3.1.1 Modification of iPP by corona discharge

One of the most effective methods of modification of iPP is its surface modification using electric discharge plasma at the atmospheric pressure (corona discharge) (Sun et al., 1999; Strobel et al., 2003). This method of modification allows preservation of the original mechanical properties of the iPP provided the optimum parameters of electric discharge have been set while the surface energy and polarity increase. In our experiments (Novák & Chodák, 1998), summarized in Figs. 1 – 8, the surface properties of iPP modified by corona discharge at varying parameters, e.g. current density, exposition time, aging, have been studied and the influence of polymer crystallinity on the surface properties has been investigated, too.

The dependence of the surface energy of the iPP modified by corona discharge on current density is represented in Fig. 1. After an initial induction period (up to 0.4 mA) a more rapid increase in surface energy was observed. As seen in Fig. 1, the increase in surface energy is by 3 - 5 $mJ.m^{-2}$ higher for the extruded iPP compared to the biaxially oriented polymer. According to Fig. 1, an increase of the surface energy was observed for biaxially oriented iPP (curve b) from 30 to 39 $mJ.m^{-2}$, while for extruded iPP (curve a) surface energy value was found to be 49 $mJ.m^{-2}$. At higher current densities (I > 0.6 mA) the difference between the surface energy values of modified extruded and biaxially oriented iPP was enhanced from 3 to 10 $mJ.m^{-2}$. Rotardation was observed for biaxially oriented iPP above the current density of 0.6 mA. On the other hand, no retardation occurred for the extruded iPP with lower crystallinity. The observed phenomena can be explained by some kind of saturation of the surface of biaxially oriented iPP by degradation products formed during modification.

The difference in the values of the surface energy between two corona discharge modified iPP results from the different degree of crystallinity. A higher degree of crystallinity of biaxially oriented iPP prevents surface oxidation. The oxidation takes place in amorphous phase and at defect spots, which are expected to be more frequent on the surface of extruded foil when compared with the surface of the biaxially oriented iPP foil.

The dependence of the total surface energy of biaxially oriented iPP as well as its dispersive component on current density is shown in Fig. 2. The surface energy of non-modified biaxially oriented iPP is very low and consists mainly of the dispersive component. These results reflect poor adhesive properties of non-modified iPP. A substantial rise of the surface energy was observed after an initial induction period (Fig. 2, curves a, b). A more intensive retardation was observed for dispersive component of the surface energy (plot b) at the current density above 0.6 mA. On the other hand, lower retardation occurred for the total surface energy (plot a). The explanation of observed dependences in Fig. 2 is similar to that for dependences in previous Fig. 1, i.e. the saturation of polymer surface by degradation products formed during the modification. The values of dispersive part of the surface energy are stabilized.

The dependence of PC of the surface energy on the current density is shown in Fig. 3 for biaxially oriented iPP modified by corona discharge. The PC of the surface energy for unmodified biaxially oriented iPP is low (0.4 mJ.m^{-2}). Modification by corona discharge results in a significant increase of the PC of the surface energy. The increase in current density leads to a linear rise of the PC of the surface energy. The value 5 mJ.m^{-2} was reached at I = 0.8 A representing 12.8 % of the value of the surface energy.

The dependence of the surface energy of iPP modified by corona discharge on exposure time defined by a speed of the foil drive is represented in Fig. 4. The increase in surface energy of both iPP is much less pronounced at higher speed of foil drive, which is inversely proportional to the exposure time of corona discharge modification. According to the Fig. 4, extruded iPP is more sensitive to the changes of speed of foil drive than biaxially oriented foil since the increase in the speed of foil drive from 4.8 to 90 m.min^{-1} leads to a decrease of the surface energy for extruded iPP from 49 to 25 mJ.m^{-2} while for biaxially oriented iPP the values change from 39 to 32.5 mJ.m^{-2}.

The concentration of hydroperoxides in the biaxially oriented iPP modified by corona discharge increased linearly with current density (Fig. 5). If the current density rises by 0.5 mA, the concentration of hydroperoxides increased by a factor 2.9. A determination of hydroperoxides groups amount as a function of corona discharge current density allows estimating the iPP hydrophilicity according to the concentration of polar groups. The dependence in Fig. 5 is in agreement with the dependence of PC of the surface energy shown in Fig. 3. From the comparison of these two dependences a linear dependence of PC of the surface energy with hydroperoxide concentration can be expected.

If the time of exposure to corona discharge of the sample of biaxially oriented iPP is shortened, the concentration of hydroperoxides decreases (Fig. 6). The effect of the increase in the speed of the foil movement at continuous modification of PP by a corona treatment is similar to the decrease of current density. The dependences in Fig. 6 are nonlinear and demonstrate the reciprocal relation between the effect of speed of foil movement, i.e. the time of corona treatment, and current density. According to Fig. 6 the hydroperoxide concentration was the same if the speed was 90 m.min^{-1} and I = 0.8 mA or if the foil moved

4.8 m.min[-1] at I = 0.3 mA. If the speed of foil drive rises from 40 m.min[-1] up to 80 m.min[-1], the concentration of hydroperoxide is only about 6 % of the value for lower speed.

The most serious problem preventing wide application of corona discharge modification of polymers is the durability of the effect, i.e. a rapid decrease of wettability during hydrophobic recovery. One explanation of the deterioration in surface properties of modified iPP has been considered to be migration of the created polar moieties into the polymer bulk. Another explanation is based on a decomposition of polar groups formed during aging. Because of hydrophobic recovery the surface energy of extruded iPP rapidly decreases during the first 24 hours (Fig. 7, plot a), then the decrease slows down and after 240 hours it levels off. Different behaviour was observed for the biaxially oriented iPP (Fig. 7, plot b). The surface energy in this case did not change significantly during storing. The value of the surface energy of biaxially oriented iPP is almost identical with the corresponding value for the foil of extruded iPP stored for a long time.

The modified layer on iPP surface consisting from macromolecules with many oxidizing groups contributes to the self-adhesive properties of iPP. The weak cohesion of this, only several micrometers thick layer, causes the adhesive joints to be less mechanically resistant. The stability of hydroperoxides at the surface was investigated by measuring the strength of auto adhesive joints (Fig. 8). At the contact of two surfaces of modified iPP foils new bonding may arise between these surfaces because of hydroperoxide decomposition. The data in Fig. 8 indicate that the work of auto adhesion increases with temperature of the formation of auto adhesive joints. This fact corresponds well with the decrease of the concentration of hydroperoxides determined after destruction of auto-adhesive joints. The concentration of hydroperoxides is almost zero at 423 K.

The practical aspect of the iPP surface modification by the corona discharge can be demonstrated on the adhesive properties of modified and unmodified iPP carpet fabrics treated by rubbing pastes based on butyl acrylate-vinyl acetate copolymer. The results are summarized in Table 1. According to Table 1 the value of mechanical work of adhesion towards PP is rising with the increase of the rubbing paste mass.

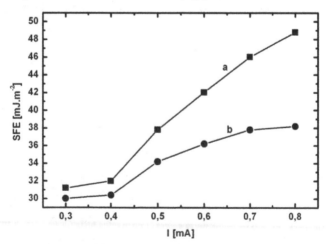

Fig. 1. Surface energy of iPP modified by corona discharge (v = 4.8 m.min[-1]) as a function of current density: a - extruded iPP, b - biaxially oriented iPP

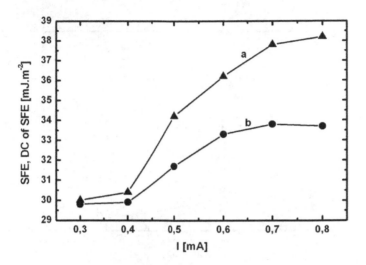

Fig. 2. Surface energy and its dispersive component as a function of current density for the biaxially oriented iPP modified (v = 4.8 m.min⁻¹) by corona discharge: a - surface energy, b - dispersive component of the surface energy

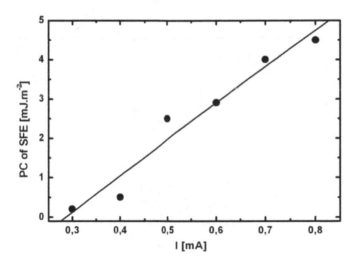

Fig. 3. PC of the surface energy as a function of current density for the biaxially oriented iPP modified (v = 4.8 m.min⁻¹) by corona discharge

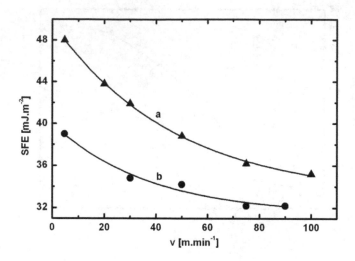

Fig. 4. Surface energy of iPP modified by corona discharge as a function of foil drive (I = 0.8 mA): a - extruded iPP, b - biaxially oriented iPP.

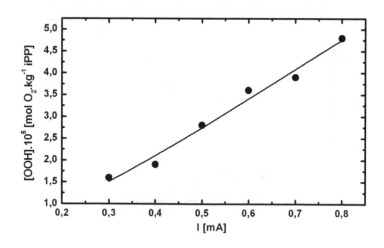

Fig. 5. Concentration of hydroperoxides in the biaxially oriented iPP modifiedby corona discharge as a function of current density (v = 4.8 m.min-1)

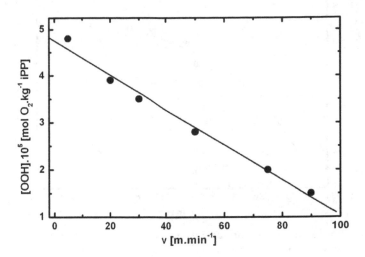

Fig. 6. Concentration of hydroperoxides in the biaxially oriented iPP modified by corona discharge as a function of foil drive speed (I = 0.8 mA)

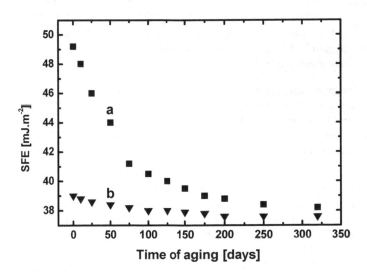

Fig. 7. Variation of the surface energy of iPP modified by corona discharge with time of aging: a – extruded PP, b – biaxially oriented PP

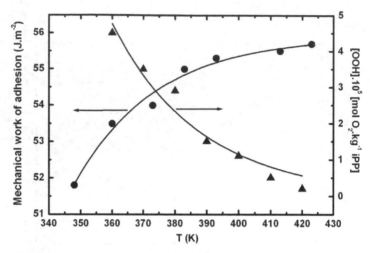

Fig. 8. Variation of auto-adhesion work and hydroperoxide concentration of iPP modified by corona discharge with the time of auto-adhesive joint formation

	Deposit of butyl acrylate-vinyl acetate copolymer (g.m⁻²)			
	50	100	130	150
	Am (J.m⁻²)			
Unmodified tissue	42	50	59	69
Modified tissue	78	84	94	102

Table 1. Mechanical work of adhesion in the adhesive joint iPP backing fabric-butyl acrylate-vinyl acetate copolymer deposit

Modification is the most important factor, since with 50 g.cm⁻² of the rubbing paste on iPP fabric the mechanical work of adhesion was higher than that for unmodified PP fabric with 150 g.cm⁻² layer of the rubbing paste. The adhesive properties given as the values of the work of adhesion improved, on average, by 65 % when compared the backing fabric based on modified PP fabrics non-modified material.

The higher efficiency of the modification for extruded iPP when compared to biaxially oriented iPP was explained by different susceptibility to oxidation due to different crystallinity. Linear dependence of either PC of the surface energy or hydroperoxide concentration of modified iPP on current density was observed. A decrease in the exposition time of iPP foil resulted in lower modification effect. The sensitivity on exposition time was more pronounced for extruded iPP. Aging of modified PP foil leads to a significant drop in surface energy during first 24 hours for extruded iPP, while for biaxially oriented iPP the decrease was small. An increase of the mechanical work of adhesion was observed if temperature of auto-adhesive joints rised. The observed decrease of hydroperoxide concentration after a destruction of the auto-adhesive joints is in accordance with the previous data.

The insufficient stability of the polarity of such a polymer surface after modification by the corona discharge brings some difficulties, i.e. due to the fact that some instable oxygenic functional groups such as peroxide or hydroperoxide groups are formed and subsequently decompose to give the more stable ketone and aldehyde groups. Because of thermodynamic preference, the surface energy of the polymer decreases due to the transfer of polar functional groups from the surface of polymer into the bulk. Owing to hydrophobic recovery following modification of iPP its polarity decreases and the original modification effect grows weak. The process of hydrophobic recovery takes place on the surface of polymer after modification and is dependent on the current density used as well as on the polymer crystallinity. The decrease in polarity of the modified polymer manifests itself by a decrease in wettability of the polymer surface with polar liquids, e. g. water, glycerol, liquid polar inks etc. It is therefore important to scrutinize the process of hydrophobic recovery of iPP modified by corona discharge plasma and to take into account this fact before further processing of the stored modified iPP films.

Variation of the surface energy and its PC with time after modification is represented in Fig. 9 for biaxially oriented iPP modified by corona discharge. During aging the hydrophilicity of the modified biaxially oriented iPP films initially dropped. The decrease in hydrophilicity (Fig. 9) manifested itself by a non-linear fall in surface energy (plot a) as well as PC of the surface energy (plot b) while the greatest fall in surface energy and its PC was found after 30 days following modification when the value of the surface energy had fallen from 39.2 to 37.0 mJm^{-2}. A greater relative decrease was observed for PC of the surface energy, which fell from 9.2 to 6.8 mJ.m^{-2} during 30 days after modification. In the course of further aging the process of iPP hydrophobization proceeds rather slow and after 360 days the surface energy and its PC reached 35.7 mJ.m^{-2} and 5.5 mJ.m^{-2}, respectively. From these values, one year after modification of the biaxially oriented iPP by the electric discharge the total surface energy decreased only a little - 9%, while the PC of the surface energy fell up to 40% when compared with that of a fresh modified sample. The measurements of the surface energy and its PC biaxially oriented iPP modified by the corona discharge show an increase of investigated surface parameters while the dispersive component of the surface energy remained practically unchanged.

Dependencies of the surface energy and its PC against long-term aging are represented in Fig.10 for extruded iPP modified by corona discharge (Novák & Florián, 2001; Novák & Florián, 2004). In comparison with biaxially oriented iPP the hydrophobic recovery of extruded iPP films subjected to modification by the electric discharge was much more obvious and the total surface energy (plot a) and its PC (plot b) fell especially in the course of the first 30 days after modification in the same manner as observed for biaxially oriented iPP. According to Fig. 10 the decrease in the hydrophilicity of extruded iPP manifested itself by a non-linear decrease in the surface energy and the PC of the surface energy. While the value of the total surface energy dropped significantly from 46 to 35 mJ.m^{-2}, the PC of the surface energy declined still more in this interval, i.e. from 15.8 to 4.6 mJ.m^{-2}. In the course of further aging of the extruded iPP the process of hydrophobic recovery of modified surface became slower and after 360 days the values of the surface energy and its PC attained 33.4 mJ.m^{-2} and 3 mJ.m^{-2}, respectively. On this basis we can state that the total surface energy of the modified extruded iPP decreased by 24% after 30 days and 27% after 360 days following modification. On the other hand, the PC of the surface energy decreased by 71% after 30 days, and by 81% after 360 days when compared with original modified sample. If we compare the values of the surface energy and its PC obtained for the biaxially oriented iPP and the extruded iPP modified by the corona discharge, we can conclude that the extruded iPP exhibits essentially higher dynamics for the hydrophobic recovery process than biaxially oriented iPP does (compare Fig. 9 a and Fig. 10). This difference is to be attributed to the different crystallinity of extruded and biaxially oriented iPP. The modification of iPP due to linking polar functional groups to the polymer chain takes place in the amorphous phase of the polymer. This is more significant in the case of extruded iPP, which has lower crystallinity when compared with biaxially oriented iPP. Owing to this lower crystallinity of modified extruded iPP a higher polymer polarity and thus a higher value of its surface energy results. Fig. 11 depicts the polar fraction for biaxially oriented iPP (plot a) and for extruded iPP (plot b) after modification by the electric discharge in the process of long-term hydrophobic recovery.

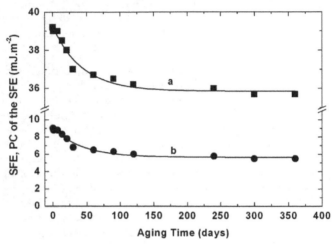

Fig. 9. Surface energy and its PC of biaxially oriented iPP modified by corona discharge during long-term hydrophobic recovery: a – surface energy, b – PC of the SFE

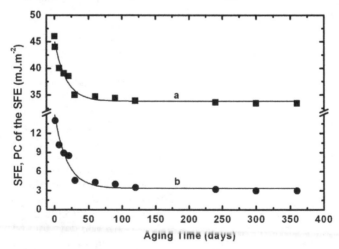

Fig. 10. Surface energy and its PC of extruded iPP modified by corona discharge during long-term hydrophobic recovery: a – surface energy, b – PC of the SFE

According to Fig.11 both relationships exhibit non-linear character and the marked decrease in the case of extruded iPP amounted to 62 % after 30 days and after 360 days following modification it was 74%. A slight decrease in hydrophilicity was observed after modification of biaxially oriented polymer, i.e. the polar fraction was reduced by 22% after 30 days and 35% after 360 days following treatment. The greatest decrease in the polar fraction of iPP modified by electric discharge was observed over 30 days after modification and during further aging the decrease was rather less for both extruded and biaxially oriented iPP. If we compare plot a with plot b in Fig.11, we can state that the decrease in the polar fraction for extruded iPP modified by the electric discharge and exposed to long-term aging was twofold when compared with that of biaxially oriented iPP.

Fig.12 presents the dependence of the mechanical work of adhesion of iPP modified by the electric discharge for an adhesive joint with polyvinyl acetate on aging the modified biaxially oriented iPP (plot a) and extruded iPP (plot b). In the course of aging of the iPP films the mechanical work of adhesion to polyvinyl acetate fell in a non-linear manner. After 360 days of aging the mechanical work of adhesion fell 26 % for modified biaxially oriented iPP when compared with original value of the mechanical work of adhesion and by 52 % for modified extruded iPP. As for the polar fraction the greatest decrease in mechanical work of adhesion to polyvinyl acetate produced by hydrophobic recovery of the surface of modified iPP was observed after 30 days following modification. During further aging a significant decrease in mechanical work of adhesion was also observed.

Because of the apparent correlation between mechanical work of adhesion and the polar fraction found for iPP modified by the electric discharge we have constructed the relationship between mechanical work of adhesion and polar fraction given in Fig. 13. The polar fraction faithfully expresses the change in polarity of the modified iPP during long-term aging. Thus Fig. 13 gives the change in mechanical work of adhesion in the course of long-term hydrophobic recovery of the surface of the modified iPP. The discussed relationship exhibits linear character and can be described by the equation:

$$A_m = 65.9 + 3.9 \cdot 10^2 \cdot x_s^p, \; r^2 = 0.99 \tag{20}$$

Because of the value of correlation coefficient found we can state that the linearity of the relationship given in Fig. 13 has been confirmed for the long-term aging of iPP after modification.

The relative decrease in the strength of adhesive joints of the modified iPP to polyvinyl acetate arising during the process of hydrophobic recovery is represented in Fig. 14 for biaxially oriented iPP (plot a) and extruded iPP (plot b) modified by the electric discharge. According to Fig. 14 the decrease in adhesive properties is significantly smaller for biaxially oriented iPP when compared with that of the extruded iPP in contradiction with assumptions based on the values found of the PC of the SFE. The decrease in values of the PC of the surface energy, the polar fraction and the mechanical work of adhesion for iPP modified by electric discharge may attributed to the successive destruction of unstable oxygenic functional groups (peroxides and hydroperoxides) arising in the initial stage of modification which give rise to the formation of more stable products. The process of hydrophobic recovery after iPP modification results from the tendency of the polymer to reduce the surface energy because of the thermodynamic preferences present and thus to rearrange the polar functional groups in the direction of polymer bulk. Because of this, the degree of surface modification of iPP by the corona discharge is, to certain extent, dependent on the crystallinity of the polymer and on time elapsed after polymer modification.

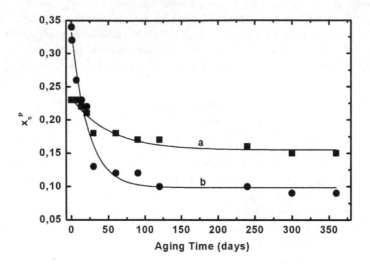

Fig. 11. Polar fraction of the iPP surface modified by corona discharge during long-term hydrophobic recovery: a – biaxially oriented iPP, b – extruded iPP

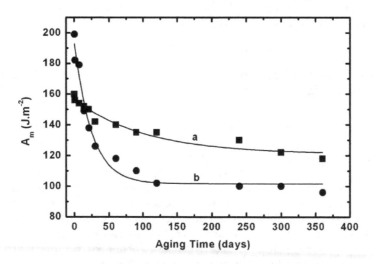

Fig. 12. Mechanical work of adhesion to polyvinyl acetate for iPP modified by corona discharge during long-term aging: a – biaxially oriented iPP, b – extruded iPP

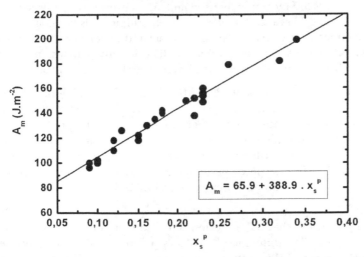

Fig. 13. Dependence of the mechanical work of adhesion to polyvinyl acetate for iPP modified by corona discharge plasma vs. polar fraction

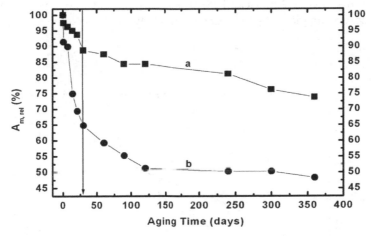

Fig. 14. Variation of the relative change in the strength of the adhesive joint modified iPP – polyvinyl acetate during long-term hydrophobic recovery: a – biaxially oriented, b – extruded iPP.

3.2 Modification by grafting

Grafting of some polar polymers (Rätzsch et al., 2002; Castell et al., 2004; Kuhn et al., 2000; Scarlatti et al., 2004; Shi et al., 2001; Pesetskii et al., 2001; Kato et al., 2003; Sulek et al., 2001; Tao et al., 2001; Flores-Gallardo et al., 2001; Yamada et al., 2003), such as acryl amide, maleic anhydride, methyl methacrylate, itaconic acid (IA) onto iPP chain, is a possibility for the increase of both the surface polarity and hydrophilicity of iPP. The free-radical polymerization methods are most widely used and inexpensive procedures for the synthesis of graft copolymers of iPP because they are relatively simple. This is a common method of increasing of iPP adhesive properties. The main chain and the branch chains in grafted iPP are usually thermodynamically incompatible. Grafting improves adhesion, dyeing, tensile strength, compatibility, thermal stability and abrasion resistance of the copolymer.

The results of surface characteristics of iPP grafted by IA (Novák & Florián, 1995; Novák & Chodák, 1995) are shown in Fig. 15. As seen in Fig. 15, plot a, the surface energy of iPP grows with the increase of concentration of grafted IA. The grafting of 5 wt. % of IA onto iPP macromolecule increased the surface energy by 19.5 % in comparison with unmodified iPP. This increase should be attributed mainly to the growth of PC of the surface energy (Fig. 15, plot b). The PC of the surface energy increased after grafting of iPP by 5 wt.% from the value 0.4 mJ.m^{-2} (non-modified polymer) to 9.5 mJ.m^{-2} obtained for iPP containing 5 wt.% of grafted IA.

The results in Fig. 16 show that the adhesion of iPP to more polar polymers as well as to metals (Fig. 17) is low. The difference between a polarity of iPP and other more polar polymer as a consequence of iPP grafting by IA was diminished. The adhesion parameters in the system iPP grafted by IA - more polar polymer increases with an increase of the IA grafted content. Grafting of iPP by 5 wt.% of IA leads to a growth of mechanical work of adhesion of iPP 6.9 times, 7.2 times and 18.1 times when considering the adhesion of iPP to cellulose acetate, polyamide and aluminum, respectively.

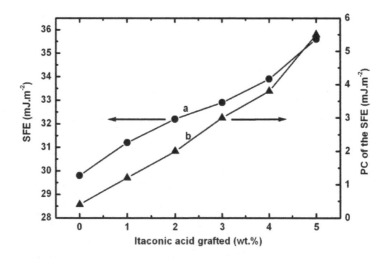

Fig. 15. Surface energy and its PC of iPP grafted by itaconic acid against itaconic acid grafted content

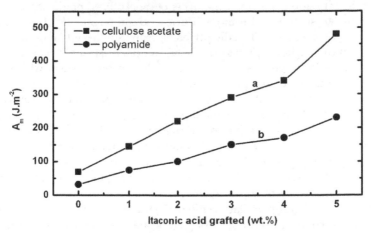

Fig. 16. Mechanical work of adhesion of iPP grafted by itaconic acid to cellulose acetate (a) and polyamide (b) against itaconic acid grafted concentration

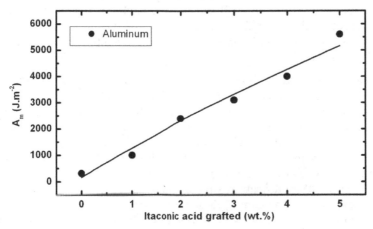

Fig. 17. Mechanical work of adhesion of iPP grafted by itaconic acid to aluminum against itaconic acid grafted concentration

3.3 Modification by oxidation

The significant groups of adhesive properties of iPP increasing present oxidation methods (Brewis & Mathieson, 2002; Chodák & Novák, 1999; Ashana et al., 1997; Novák, 1996; Yang et al., 2003; Nie et al., 2000; Novák & Pollák, 1994; Novák, 1996; Dibyendu et al.,1998; Vasconcellos et al.,1997; Novák & Chodák, 2001). As a result of formation of carbonyl groups in the course of iPP modification the adhesive properties of polymers are increased. Using a strong oxidation agent and higher temperatures of modification a partial decomposition of the polymer can occur when the bulk unmodified zones of polymer reach the surface. On spherollitic surfaces of iPP chromic acid preferentially etches the less ordered (amorphous) regions between the arms of spherullites, producing ca 10 μm deep cavities. On lamellar surfaces, chromo sulfuric acid preferentially attacks interlamelar regions, producing deep cavities. The deeply etched spherullitic and lamellar surfaces have numerous mechanical anchoring sites for metal to adhere strongly. For the increase of adhesion of iPP functional groups =C=O and -SO$_3$H are responsible, which are formed during the modification and are present in the near-surface layer. In iPP modified by chromo sulfuric acid there is also a high concentration of carboxyl groups. After the modification of iPP by chromo sulfuric acid a very rough surface is formed and that the roughness itself can cause a good adhesion of treated iPP. A different opinion on the better adhesion properties after modification by chromo sulfuric mixture considers a main reason for the better adhesion properties to be the higher polarity of pre-treated iPP, while the higher roughness to be a secondary importance contributing the higher adhesion only. For the modified iPP similar correlations were found between the degree of oxidation or concentration of carboxyl group and adhesion. From the results of the studies on the surface energy of iPP modified by chromo sulfuric acid it follows that after initial fast increase, during 1 - 2 minutes of modification, the value of the surface energy levels off. The most suitable temperature for modification of iPP by chromo sulfuric acid is 70 ºC, at which the degradation of polymer is still not extensive. In the measurement of mechanical work of adhesion of iPP modified by chromo sulfuric acid in a joint with polyvinyl acetate the result was six times larger relatively to the untreated iPP. This result is similar to that for the modification by corona discharge besides the fact that the modification by chromo sulfuric acid has a lasting character.

An efficient method of modification of iPP from the point of view of enhancement of adhesion properties is a modification by vapors of chromyl chloride (Novák, 1996; Novák & Pollák, 1994). In contrast to the modification by chromo sulfuric acid a considerably longer time of modification is needed (40 - 60 min.) and the treatment is performed in the temperature range of 30 - 60 ºC. The study shows few times higher concentration of carbonyl groups in comparison to treatment by chromo sulfuric acid, while Cr^{VI} was reduced to lower oxidation degrees up to Cr^{IV}. In the study of a dependence of PC of the surface energy on time of modification a maximum was observed and it was confirmed at the same conditions also for the similar dependence of concentration Cr^{IV} by ESR method. In the course of modification of iPP by chromyl chloride at higher degrees of oxidation a degradation of iPP (linear decrease of molar mass with the time of modification) was observed. The modification mechanism of chemical etching by oxidizing agents consists in the abstraction of the hydrogen atoms from the polymer backbone and their replacement with polar groups. These polar groups introduced on the iPP surface by pretreatment should increase the surface energy and enhance the wetting. The surface pretreatment by some oxidizers has also another aim - to remove all weak boundary layers on the polymer surface e.g. stabilizers and other additives, which are responding for the weak adhesive bonds. The oxidizing chemical methods of iPP modification are efficient and in addition to this fact they give a more durable result than modification by corona discharge.

The results of the investigation of surface properties of iPP pretreated by chromyl chloride and/or chromo sulfuric acid are presented in Figs. 18 - 26.

Variation of the surface energy of iPP modified by chromyl chloride with modification time is represented in Fig. 18. It follows that the surface energy non-linearly increases with modification time. However, it becomes stable after a certain period of time. The lower the temperature of modification, the later the stable state appears. For instance, the stable state of the relationship described in Fig. 18 appeared after 30 minutes if the modification was performed at 333 K and after 60 minutes if the modification was performed at 303 K. The stabilization of the values of the surface energy and the shift towards longer time due to the decrease in the temperature of oxidation (Fig. 18) can be explained by stepwise saturation of the polymer surface with polar (mainly carbonyl) groups, which originate from modification of iPP by chromyl chloride. The increase in surface energy of iPP modified by chromyl chloride (Fig. 18, plots a, b, c) was 36 % for the temperature of 333 K, 28 % for 313 K and 13 % for 303 K with respect to non-modified polymer.

The dependence of the surface energy and its PC against time of oxidation is represented in Fig. 19 for iPP modified by chromyl chloride. According to Fig. 19, the PC of the surface energy of modified iPP increases during the first 30 - 40 min. of modification, after this period the value of PC decreases. The reaching the maximum of PC of the surface energy may be probably ascribed to degradation processes occuring at the modified polymeric surface, which could decrease the polarity of the surface. After a certain period of time the degradation products accumulate at the surface of iPP reducing the polarity, which brings a decrease in surface energy of polymer.

In iPP modified by chromyl chloride the concentration of carbonyl groups was determined (Fig. 20). The concentration of the carbonyl groups increased with the time of oxidation and reaches a constant value after 30 - 60 min., depending on the temperature of modification. Plots a, b, c in Fig. 20 correspond to the temperatures 333 K, 313 K and 303 K, respectively, exhibit analogous course like the plots of the surface energy in Fig. 24. Change the temperature of modification from 303 to 333 K, the concentration of the surface carbonyl groups increases by the factor 4.4. Saturation of the surface with carbonyl groups is probably the reason for the stabilization of the carbonyl groups concentration after a given time.

Determination of the reaction rate constant of iPP oxidation by chromyl chloride showed a strong dependence of the reaction rate on the temperature (Fig. 21). The growth of the temperature from 303 K to 333 K caused an increase of the reaction rate by the factor 1.9.

The results of the CrVI to CrIV reduction in iPP modified by chromyl chloride obtained using electron spin resonance are represented in Fig. 22. It can be seen that the maximum Cr^{IV} concentration at 333 K is reached after 40 min. of oxidation. As it can be seen, the maximum concentration of Cr^{IV} was reached after 40 minutes of iPP oxidation at the temperature of 333 K and the curves in Fig. 22 and Fig. 20 are in good correlation.

The dependence of the surface energy and its PC vs. concentration of C=O groups is shown in Fig. 23. The PC of the surface energy reaches a maximum at the C=O groups concentration equal to 1.6×10^{-2} mol $O^2.kg^{-1}$ iPP. The following decrease in the C=O groups concentration is probably caused by an increased amount of the degradation products thus resulting in the decrease of the value of the surface energy.

The degree of the iPP degradation during the modification by chromyl chloride was estimated from the determination of its molecular weights. Like it can be seen from Fig. 24, the molecular weight of iPP decreases linearly with the time of modification, the slope of the curves is increasing with increased temperature. After 60 min. of modification the molecular weight decrease of iPP is about 10 % for T = 303 K, 26 % for T = 313 K and 42 % for T = 333 K, in comparison with the unmodified iPP.

The decrease of the surface energy with time after iPP modification is described in Fig. 25. As it follows from this figure, the iPP surface after modification by chromyl chloride has a relatively good stability. The decrease of the surface energy value 120 hours after finishing modification is close to 7%, while this decrease reaches 30 – 40 % in 24 hours after modification of iPP by corona discharge.

The mechanical work of adhesion of iPP modified by chromyl chloride and chromo sulfuric acid vs. time of modification is illustrated in Fig. 26. It was found that the mechanical work of adhesion is 36 % higher in the case of chromyl chloride compared to the modification with chromo sulfuric acid. The value of mechanical work of adhesion increased rapidly, in a short time (2 min.) reached the maximum value and soon afterwards assumes a constant value.

It has been found that the surface energy of iPP pretreated by chromyl chloride versus modification time rapidly increases and assumes a constant value. Equal character was observed for the relationship between free surface energy and carbonyl group concentration of modified polymer. After a certain time interval subsequent to modification the decrease in surface energy of treated iPP is negligible. A maximum of the relationship between PC of the surface energy and modification time was revealed.

A quite often used method of modification of iPP is the modification by UV light. In analogy to the treatment by the electric discharge the effect of UV light is exhibited in the air by the formation of hydroperoxides which are consequently transformed by photolysis to ketones. It is known that by the decomposition of hydroperoxides, OOH radicals are formed which can lead to formation of ester, carboxyl and lacton groups and thus enhance the adhesion of iPP. A grafting of polar monomers such as N-vynilpyrrolidon or methacrylic acid to iPP modified by UV light can be used. Enhancement of adhesion was observed and could be effectively expressed by a decrease of contact angle of redistilled water on the surface of polymer from 94 deg to 35 – 60 deg. For modification of iPP is very important to know the effect of the intensity and the distance of the UV source from the treated polymeric surface. A very efficient method of adhesion enhancement of iPP is the modification by phosphorus trichloride with participation of the UV light. By the decomposition of phosphorus trichloride active chlorine is formed which reacts with iPP. This is also accompanied at the same time by the oxidation by ozone formed from oxygen in the air. A significant enhancement of adhesion was observed without a considerable decrease of the film transparency.

The inferior adhesive properties of polyolefins result in many serious problems especially if gluing or printing on these materials is considered. The effective surface modification of polyolefin should lead to a formation of a very thin surface layer with the thickness of several micrometers without affecting the bulk properties of the material. The fine layer of modified polymer on the surface of polyolefin should contain sufficient concentration of the polar moieties leading to an increase of the surface energy of polymer.

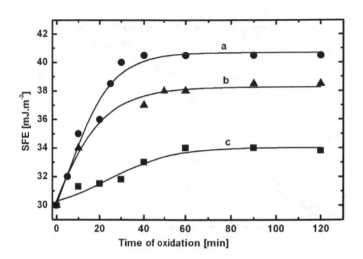

Fig. 18. Surface energy of iPP modified by chromyl chloride against modification time:
a - T = 333 K, b - T = 313 K, c - T = 303 K

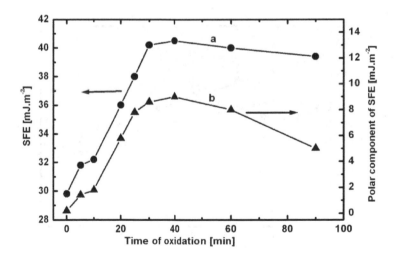

Fig. 19. Surface energy and its PC against time of modification for iPP modified by chromyl
chloride: a - surface energy, b - PC of the surface energy.

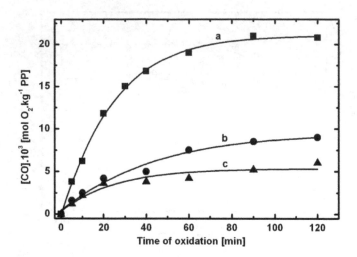

Fig. 20. Concentration of carbonyl groups of iPP modified by chromyl chloride against time of modification: a - T = 333 K, b - T = 313 K, c - T = 303 K

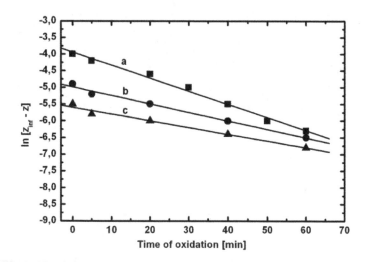

Fig. 21. Determination of the reaction rate constant of iPP modified by chromyl chloride at different temperatures: a – 333 K, b - 313 K, c - 303 K

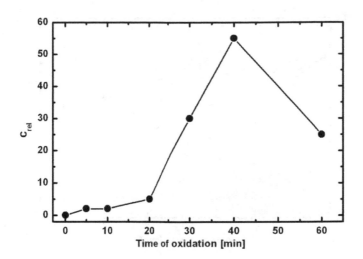

Fig. 22. ESR measurements of Cr^{IV} concentration of iPP modified by chromyl chloride against time of modification (T = 333 K)

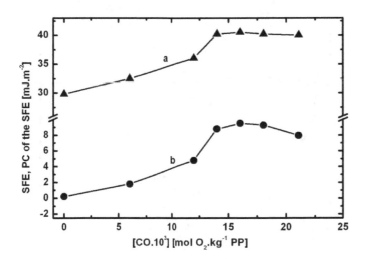

Fig. 23. Surface energy and its PC of iPP modified by chromyl chloride against carbonyl groups concentration: a – surface energy, b - PC of the surface energy

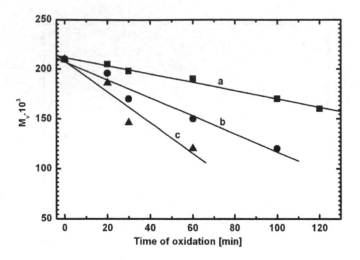

Fig. 24. Figure 24 Molar weight of iPP modified by chromyl chloride vs. time of modification: a – T = 303 K, b – T = 313 K, c - T = 333 K

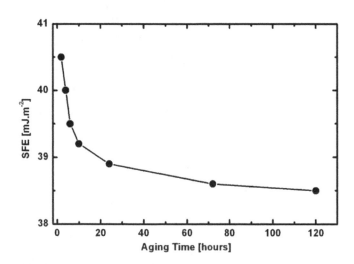

Fig. 25. Surface energy of iPP modified by chromyl chloride vs. time elapsed from modification

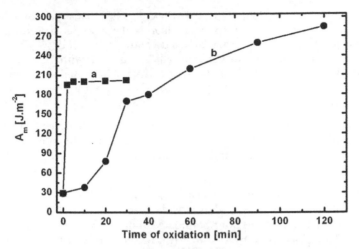

Fig. 26. Mechanical work of adhesion of iPP modified by chromyl chloride to polyvinyl acetate against time of modification: a – chromo sulfuric acid, b – chromyl chloride

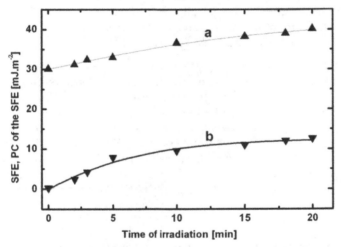

Fig. 27. Variation of the surface energy and its PC of iPP pre-treated by UV/POCl₃ with the time of UV exposure (d_{UV} = 50 mm): a – surface energy, b - PC of the surface energy

3.4 Modification by halogenization

Surface modification of iPP in vapours of halogen compounds (Novák & Chodák, 2001; Novák & Chodák, 2001; Kharitonov, 2000; Novák& Chodák, 2001; Hruska & Lepot, 2000; Carstens et al., 2000) under UV irradiation represents an efficient method for the increase of the adhesive properties. The presented method is based on the influence of the UV radiation and vapours of phosphoryl chloride on the iPP surface. UV-irradiation results in a faster decomposition of the halogen compound and leads to the UV-oxidation of the surface of polymer. Surface modification of iPP vapours of phosphoryl chloride under UV irradiation

(UV/POCl₃) (Novák & Chodák, 2001; Novák & Chodák, 1999) is a suitable method for the increase of adhesive properties. Phosphoryl chloride behaves as a sensitizer, which decomposes under the effect of UV irradiation. Its decomposition is followed by a formation of active chlorine, which takes part in the free radical reactions with PO macromolecules leading to a formation of $-POCl_2$ groups attached to the polymer chains. At the same time the reaction with oxygen initiated by UV irradiation takes place resulting in a generation of polar oxygen-containing sites.

The surface and adhesive properties of iPP modified by UV radiation in the presence of POCl₃ have been studied. Modification of iPP by UV/POCl₃ results in an increase in surface energy of polymer as well as in an improvement of the mechanical work of adhesion of modified iPP to more polar polymers.

Fig. 27 illustrates a non-linear increase of the surface energy (curve a) and its PC (curve b) for iPP modified by UV/POCl₃ in the dependence on the time of modification at the UV source distance of 50 mm. The surface energy for non-modified polymer was 30.1 mJ.m⁻². The surface energy and its PC of iPP pre-treated by UV/POCl₃ increased non-linearly with time up to the value 40.1 mJ.m⁻² after 20 minutes. The PC of the surface energy of modified iPP increased after 20 min. of modification from 0.25 mJ.m⁻² up to 12.6 mJ.m⁻². The time-dependence of the PC of the surface energy (Fig. 27, curve b) leveled off after an initial increase enduring about 10 min. As expected, the strength of adhesive joints of modified iPP increases in accordance with the degree of hydrophilicity of modified polymer. Thus, adhesive properties of polymer are strongly related to the values of the PC of the surface energy.

The surface energy of iPP modified by UV/POCl₃ was essentially higher in comparison with non-modified polymer. A non-linear increase of the surface energy and its PC against time of UV exposure was observed. The pre-treatment of iPP by UV/POCl₃ method leads to a considerable growth in the surface and adhesive properties of the polymer and it depends on the intensity of the UV radiation as well as on the time of UV exposure. The values of the strength of adhesive joint to polyvinyl acetate of UV/POCl₃ modified iPP correspond with the measured values of the PC of the surface energy. The efficiency of iPP modification by UV/POCl₃ substantially increased by diminishing the distance of UV radiation source from the surface of polymer or by increasing the intensity of UV source.

4. Conclusion

The selection and application of modification procedure for the enhancement of adhesion properties of iPP depends mostly on the specific demands required on the adhesive joints in practice. When deciding about suitability of the modification method a detailed knowledge about physical and chemical changes undergoing on the surface, experimental demands and also aggressiveness of the reagents employed is needed. Based on this point of view the most practical surface treatments are the flame and electric discharge treatments which enables a continual modification during the processing, for instance in printing. This modification meets also the demand on a fast processing of the treated surface. On the other hand some very effective modification methods such as the modification by chromyl chloride vapours or phosphorus trichloride with UV light application are less popular because of implausibility of the modification reagents, in spite of the lasting effect of the modification. The surface modification of iPP is observed in a very narrow layer not exceeding thickness of 10 μm and is easily destroyed by heat or friction by which the effect of modification disappears. For practical purposes according to this reason, there is also suitable method of modification based on the modification of polymer in bulk consisting of the addition of polar low-molecular compound or a polymer respectively. As a matter of fact, the most efficient method of modification is grafting, e.g. by maleic anhydride, acrylic or itaconic acid, as well as the modification by block copolymers of the A-B-C type where A is a polar monomer and B is non-polar one. This modification is now intensively studied because it enables us to obtain good properties of iPP without significant deterioration of its physico-mechanical properties.

5. Acknowledgement

The authors are grateful to Slovak Grant Agency VEGA (grants No. 2/0185/10 and 1/0581/12), Slovak Research and Development Agency (grants APVV 51-010405 and APVV-0742-11).

6. References

Ashana, H.; Erickson, B. L. & Drzal, L. T. (1997). Sulfonation of polymer surfaces. 2. Chemical changes on polypropylene and polystyrene surfaces after gas phase sulfonation. *J. Adhes. Sci. Technol.*, Vol. 11, pp. 1269-1288, ISSN 0169-4243.

Brewis, D. & Mathieson, I. (2002). M. Adhesion and bonding to polyolefins, *Rapra Review Reports: Shrewsburry*, pp. 10-14, ISBN 1-85957-323-1.

Castell, P.; Wouters, M.; De With, G.; Fischer, H. & Huijs, F. (2004). Surface modification of polypropylene by photoinitiators: Improvement of adhesion and wettability. *J. Appl. Polym. Sci.*, Vol. 92, pp. 2341-2350, ISSN 1096-4628.

Carstens, P. A. B.; Marais, S. A. & Thompson. C. J. (2000). Improved and novel surface fluorinated products. *J. Fluor. Chem.*, Vol. 104, pp. 97-107, ISSN 0022-1139.

Chodák, I. & Novák, I. (1999). Surface modification of polypropylene by additives. An A-Z Reference. *Karger-Kocsis, J.; ed.; Kluwer Acad. Publ.: Dordrecht*, pp. 790-793, ISSN 2080-8259.

Denes, F. S. & Manolache, S. (2004). Macromolecular plasma-chemistry: an emerging field of polymer science, *Progr. Polym. Sci.* Vol. 29, pp. 815 – 885, ISSN 0079-6700.

Dibyendu, S. B.; Ghosh, S. N. & Maiti, S. (1998). Surafce modification and evaluation of polyehylene film. *Eur. Polym. J.*, Vol. 34, pp. 855-861, ISSN 0014-3057.

Flores-Gallardo, S. G.; Sanchez-Valdez, S. & De Valle L. F. R. (2001). Polypropylene/polypropylene–grafted acrylic acid blends for multilayer films: preparation and characterization. *J. Appl. Polym. Sci.*, Vol. 79, pp. 1497-1505, ISSN 1096-4628.

Guimond, S.; Radu, I.; Cyeremusykin, G.; Carlsson, D. J. & Wertheimer, M. R. (2002). Biaxially oriented polypropylene [BOPP] surface modification by nitrogen atmospheric pressure glow discharge [APDD] and by air corona. *Plasmas Polymers*, Vol. 7, pp. 71-88, ISSN 1084-0184.

Hruska, Z. & Lepot, X. (2000). Ageing of oxyfluorinated polypropylene surface: Evolution of the acid-base surface characterisitics with time. *J. Fluor. Chem.*, Vol. 105, pp. 87-93, ISSN 0022-1139.

Kinloch, A.J. (1987). Surface pretreatments. In: *Adhesion and Adhesives,* pp. 101-170,Chapman & Hall: London, ISBN 0-412-27440-X.

Kolluri, O. S. (1994). Application of plasma technology for improved adhesion of materials. In: *Handbook of adhesive technology,* pp. 35-46, Pizzi, A.; Mittal, K. L; eds.; Marcel Dekker,: New York, ISBN 0-8247-8974-1.

Kim, K. S.; Yun, Y. I.; Kim, D. H.; Ryu, C. M. & Park, C.E. (2002). Inhibition of aging in plasma-Treated high-density polyethylene. *J. Adhes. Sci. Technol.,* Vol. 16, pp. 1155-1169, ISSN 0169-4243.

Kim, B. K.; Kim, K. S. & Park, C. E. (2002). Improvement of wettabilitty and reduction of aging effect by plasma treatment of low-density polyethylene with argon and oxygen mixtures. *J. Adhes. Sci. Technol.,* Vol. 16, pp. 509-521, ISSN 0169-4243.

Kuhn, G.; Ghode, A.; Weidner, S.; Retzko, I.; Unger, W. E. S. & Friedrich, J. F. (2000). Chemically well-defined surface functionalization of polyethylene and polypropylene by pulsed plasma modification followed by grafting of molecules. In: *Polymer surface modification: Relevance to adhesion,* Vol. 2, pp. 45-64, Mittal, K. L. Ed., VSP: Utrecht, ISBN-10 9067643270.

Kato, K.; Uchida, E., Kang, E. T.; Uyama, Y. & Ikada, Y. (2003). Polymer surface with graft chains. *Progr. Polym. Sci.,* Vol. 28, pp. 209-259, ISSN 0079-6700.

Kharitonov, A. P. (2000). Practical applications of the direct fluorination of polymers. *J. Fluor. Chem.,* Vol. 103, pp. 123-127, ISSN 0022-1139,

Lei, J. X.; Li, Q. M.; He, G. J. & Lin, X. H. (2000). Surface graft copolymerization of acryl amide onto BOPP film through corona discharge. *Acta Chim. Sinica,* Vol. 58, pp. 598-600, ISSN 0567-7351.

Mikula, M.; Jakubíková, Z. & Záhoranová, A. (2003). Surface and adhesion changes of atmospheric barrier discharge treated polypropylene in air and nitrogen. *J. Adhes. Sci. Technol.,* Vol. 17, pp. 2097-2110 ISSN 0169-4243.

Moosheimer, U. & Bichler, C. (1999). Plasma pretreatment of polymer films as a key issue for high food packaging. *Surf. Coat. Technol.,* Vol. 116-119, pp. 812-819, ISSN 0257-8972.

Nie, H. Y.; Walzak, M. J. & McIntyre, N. S. (2000). Atomic force study of UV/ozone treated polypropylene film. In: *Polymer surface modification: Relevance to adhesion,* Vol. 2, pp. 377-392, Mittal, K. L. Ed., VSP: Utrecht, ISBN-10 9067643270.

Noeske, M.; Degenhardt, J.; Strudthoff, S. & Lommatzsch, U. (2004) Plasma jet treatment of five polymers at atmospheric pressure: surface modifications and the relevance for adhesion. *Intern. J. Adhes. Adhesives.,* Vol. 24, pp. 171-177, ISSN 0169-4243.

Novák, I.; Krupa, I. & Luyt, A. S. (2004) Modification of the polarity of isotactic polypropylene through blending with oxidized paraffin wax. *J. Appl. Polym. Sci.,* Vol. 94, pp. 529-533, ISSN 1096-4628.

Novák, I. & Florián, S. (2001). Influence of aging on adhesive properties polypropylene modified by discharge plasma. *Polym. Intern.,* Vol. 50, pp. 49-52, ISSN 0959-8103.

Novák, I. & Florián, S. (2004). Investigation of long-term hydrophobic recovery of plasma modified polypropylene. *J. Mater. Sci.,* Vol. 39, pp. 2033-2036, ISSN 0022-2461.

Novák, I. & Chodák, I. (1998). Adhesion of polypropylene modified by corona discharge. *Angew. Makromol. Chem.,* Vol. 260, pp. 47-51, ISSN 0003-3146.

Novák, I. & Florián, S. (1995). Adhesive properties of polypropylene modified by electric discharge. *J. Mater. Sci. Lett.,* Vol. 14, pp. 1021-1022, ISSN 0261-8028.

Novák, I. & Chodák, I. (1995). Effect of grafting on polypropylene adhesive characteristics. *J. Mater. Sci. Let.*, Vol. 14, pp. 1298-1299, ISSN 0261-8028.

Novák, I. (1996). Effect of surface pretreatment on wettability of polypropylene. *J. Mater. Sci. Lett.*, Vol. 15, pp. 1137-1138, ISSN 0261-8028.

Novák, I. & Pollák, V. (1994). Surface modification of polypropylene by chromylchloride. *Angew. Makromol. Chem.*, Vol. 220, pp. 189-197, ISSN 0003-3146.

Novák, I. (1996). Surface properties of polypropylene modified by chromo suphuric acid. *J. Mater. Sci. Let.* 1996, Vol. 15, pp. 693-694, ISSN 0261-8028.

Novák, I. (1995). Increasing of the adhesive properties of polypropylene by surface modification. *Angew. Makromol. Chem.*, Vol. 231, pp. 69-77, ISSN 0003-3146.

Novák, I. & Chodák, I. (2001). Adhesive behavior of UV sensitizer modified low-density polyethylene. *J. Macromol. Sci. - Pure Appl. Chem.*, Vol. 38, pp. 11-18, ISSN 1060-1325.

Novák, I. & Chodák, I. (2001). Adhesive behavior of UV pre-treated polyolefins. In: *Thermoplastics-based blends and composites*, pp. 341-348, Macromol. Symp. Chodák, I.; Lacik, I.; eds., ISSN 1022-1360.

Novák, I. & Chodák, I. (2001). Effect of polypropylene UV modification on adhesion to polar polymers. *Petrol. Coal.*, Vol. 43, 27-28, ISSN 1337-7027.

OHare, L.A.; Leadley, S. & Parbhoo, B. (2002). Surface physicochemistry of corona-discharge treated polypropylene film. *Surf. Interface Anal.*, Vol. 333, pp. 335-342, ISSN 0142-2421.

Pocius, A. V. (1997). The surface preparation of adherends for adhesive bonding. In: Adhesion and adhesives technology. In: *An introduction*, pp. 148-182, Hanser Publ.: Munich, ISBN 3-446-21731-2.

Poncin-Epaillard, F. (2002). Characterization of CO2 plasma and interactions with polypropylene film. *Plasmas Polymers*, Vol. 7, pp. 1-17, ISSN 1084-0184.

Pesetskii, S.S.; Jurkowski, B.; Krivoguz, Y. M. & Kelar, K. (2001). Free-radical grafting of itaconic acid onto LDPE by reactive extrusion: I. Efect of initiator solubility. *Polymer*, Vol. 42, pp. 469-475, ISSN 0032-3861.

Ráhel, J.; Šimor, M.; Černák, M.; Štefečka, M.; Imahori, Y. & Kando, M. (2003). Hydrophilization of polypropylene nonwoven fabric using surface barrier discharge. *Surf. Coat. Technol.*, Vol. 169, pp. 604-608, ISSN 0257-8972.

Rätzsch, M.; Arnold, M.; Borsig, E.; Bucka, H. & Reichelt, N. (2002). Radical reactions on polypropylene in the solid state. *Progr. Polym. Sci.*, Vol. 27, pp. 1195-1282, ISSN 0079-6700.

Scarlatti, V.; Oehr, C.; Diegelmann, C. & Lobman, P. (2004). Influence of plasma functionalization of polypropylene with acrylic acid on the nucleation of $CaCO_3$. *Plasma Processes Polym.*, Vol. 1, pp. 51-56, ISSN 1612-8850.

Shenton, M. J.; Lovell-Hoare, M. C. & Stevens, G. C. (2001) Adhesion enhancement of polymer surfaces by atmospheric plasma treatment, *J. Phys. D:*, Vol. 34, pp. 2754-2760, ISSN 0022-3727.

Shi, D.; Yang, J. ; Yao, Z.; Wang, Y., Huang, H.; Jing, W.; Yin, J. & Costa, G. (2001). Functionalization of isotactic polypropylene with maleic anhydride by reactive extrusion: mechanism of melt grafting. *Polymer*, Vol. 42, pp. 5549-5557, ISSN 0032-3861.

Strobel, M.; Jones, V.; Lyons, C.S.; Ulsh, M., Kushner, M. J.; Dorai, R. & Branch, M. C. (2003) A comparison of corona-treated and flame-treated polypropylene films. *Plasmas Polymers*, Vol. 8, pp. 61-95, ISSN 1084-0184.

Strobel, M.; Jones, V.; Lyons, C. S.; Ulsh, M.; Kushner, M. J.; Dorai, R. & Branch, M. C. (2003). A comparison of corona-treated and flame-treated polypropylene films. *Plasmas Polymers*, Vol. 8, pp. 61-95, ISSN 1084-0184.

Schultz, J. & Nardin, M. (1994). Theories and mechanism of adhesion. In: *Handbook of adhesive technology*, pp. 19-34, Pizzi, A.; Mittal, K. L; eds.; Marcel Dekker, New York, ISBN 10 0824789741.

Sun, C.; Zhang, D. & Wadsworth, L. C. (1999). Corona treatment of polyolefin films – A rewiev. *Adv. Polym. Technol.*, Vol. 18, pp. 171-180, ISSN 1098-2329.

Sulek, P.; Knaus, S. & Liska, R. (2001). Grafting of functional maleimides onto oligo- and polyolefins. *Macromol. Symp.*, Vol. 176, pp. 155-165, ISSN 1521-3900.

Štefečka, M.; Kando, M.; Matsuo, H.; Nakashima, Y.; Koyanagi, M.; Kamiya, T. & Černák, M. (2004). Electromagnetic shielding efficiency of plasma treated and electroless metal plated polypropylene nonwoven fabrics. *J. Mater. Sci.*, Vol. 39, pp. 2215-2217, ISSN 0022-2461.

Štefečka, M.; Kando, M.; Černák, M.; Korzec, D.; Finantu-Dinu, E. G.; Dinu, G. L. & Engemann, J. (2003). Spatial distribution of surface treatment efficiency in coplanar barrier discharge operated with oxygen-nitrogen gas mixtures. *Surf. Coat. Technol.*, Vol. 174, pp. 553-558, ISSN 0257-8972.

Tao, G.; Gong, A.; Lu, J.; Sue, H. J. & Bergbreiter, D. E. (2001). Surface functionalized polypropylene: synthesis, characterization and adhesion properties. *Macromolecules*, Vol. 34, pp. 7672-7679, ISSN 0024-9297.

Vasconcellos, A. S.; Oliveira, J. A. P. & Baumhardt-Neto, R. (1997). Adhesion of polypropylene treated with nitric and sulfuric acid. *Eur. Polym. J.*, Vol. 33, pp. 1731-1734, ISSN: 0014-3057.

Yalizis, A.; Pirzada, S. A. & Decker, W. (2000). A novel atmospheric plasma system for polymer surface tretment. In: *Polymer surface modification: Relevance to adhesion*, pp. 65-76, Mittal, K. L. Ed., VSP: Utrecht, ISBN-10: 9067643270.

Yamada, K.; Takeda, S. & Hirata, M. (2003). Auto-hesion of polyethylene plates by the photo induced grafting of methacryl amide. *A.C.S. Symp. Ser.*, Vol. 847, pp. 511-521, ISSN: 0097-6156.

Yang, P.; Deng, J. Y. & Yang, W. T. (2003). Confined photo-catalytic oxidation: a fast surface hydrophilic modification method for polymeric materials. *Polymer*, Vol. 44, pp. 7157-7164, ISSN: 0032-3861.

9

Use of Nonwoven Polypropylene Covers in Early Crop Potato Culture

Wanda Wadas
Siedlce University of Natural Sciences and Humanities
Poland

1. Introduction

In view of the high prices obtained for new potatoes, their production is more profitable than other ways of potato usage. Early potato production is associated with considerable risk producers have to take due to high yield variability over years and rapid price decrease as a result of increased supply. The success of potato production for an early crop is dependent to the higher extent on the weather conditions in the initial period of plants vegetation, especially temperature (Nishibe et al., 1989; Sale, 1979). Too low soil temperature retards the emergence and inhibits the initial plant growth. High income obtained from early potato production is possible under conditions assuring early tuber setting and rapid gain of tuber yield, and its marketing when the price is highest. In case of early potato production, its location in regions where vegetation begin early is of high importance. In the regions with delayed vegetation, the crop of new potatoes can be forced by applying perforated polyethylene film or nonwoven polypropylene covers directly on the planted field. The favourable microclimate under the cover faciliates the emergence and futher plant growth and development in the period with less favourable weather conditions for early potatoes (Hamouz et al., 2006; Jenkins & Gillison, 1995; Michaud et al., 1990; Wadas & Kosterna, 2007a). The earlier plant emergence resulted in a more extensive ground cover during early growth and the higher leaf area index (LAI). The growth duration and leaf area determine the amount of solar radiation intercepted by the canopy and influences on the extent of photosynthesis, evaporation, transpiration and final dry matter yield (Gordon et al., 1997; Nelson & Jenkins, 1990). In Europe the covers that were most frequently used in the early crop potato culture were those of perforated polyethylene film (Friessleben, 1984; Hamouz & Rybáček, 1988; Jenkins & Gillison, 1995; Lang, 1984). Widespread availability on the market of nonwoven polypropylene and extensive promotion of these covers have resulted in a frequently reduced perforated polyethylene film with nonwoven polypropylene in the field cultivation of earlies. Polypropylene covers were introduced into agriculture on a large scale in the 1990s. Globally, they are commonly known under the name of nonwoven fabrics (Cholakov & Nacheva, 2009). This material is light, water permeable, transparent and airy. The weight of 1 m^2 of nonwoven polypropylene fabric is about 2.5 times lower than the weight of perforated polyethylene film, therefore, the covering does not cause any hinder for plant developing.

2. Field microclimate under the polypropylene cover

The environmental factor which restricts early potato planting to the highest extent is soil temperature (Nishibe et al., 1989; Sale, 1979). Presprouting seed-potatoes can be planted when soil temperature at the depth of 5-10 cm is maintained at the level of 5-6°C for several subsequent days. Earlier planting into insufficiently heated soil is risky, since even short but excessive over-cooling of young plants strongly weakens their further growth and delays obtaining marketable yield.The unfavourable effect of low temperatures in the early period of potato growth can be reduced by applying nonwoven polypropylene cover. The study carried out in the Czech Republic, Germany and Poland showed that soil temperature at a depth of 5 cm under nonwoven polypropylene was higher by 1-2°C, and at the depth of 10 cm by 2-3°C than the temperature of uncovered soil, while the air temperature at the ground was higher by 2.0°C (Demmler 1998; Hamouz et al., 2006; Lutomirska & Szutkowska, 1999; Prośba-Białczyk & Mydlarski, 1998; Wadas & Kosterna, 2007a). Soil temperature under nonwoven polypropylene was on average 1-2°C lower than under perforated polyethylene film. The difference in temperatures in afternoon hours reached even 5°C (Figure 1). In Bulgaria, soil temperature at the depth of 5 cm under nonwoven polypropylene at 8.00 a.m. was higher by 0.4-2.4°C, and at 2 p.m. on sunny days, even by 4.6°C (Cholakov & Nacheva, 2009). Higher soil temperature, as well as isolation by means of the nonwoven polypropylene cover from a relative drop in air temperature at night creates a more favourable microclimate for plant growth. According to Bizer (1997) and Dvořák et al. (2004), nonwoven polypropylene cover creates a favourable microclimate for potato emergence and growth, even when the temperature at the ground drops to -7°C. Soil temperature at the depth of 10 cm was by then almost 3°C higher (and above 0°C) in comparison to the uncovered field.

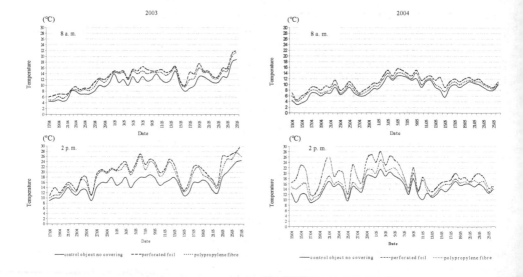

Fig. 1. Soil temperature at depth of 10 cm [°C] depending on the kind of cover at 8[00] a.m. and 2[00] p.m. in 2003 – 2004 (Wadas & Kosterna, 2007a)

3. Growth and development of plants

Profitable yield of new potatoes after 60 days from planting can already be obtained when the period between planting and plant emergence lasts 15-21 days, from plant emergence to the end of tuber seeting – 19-24 days, and the period of yield accumulation lasts a minimum 20 days (Kubiak & Gaziński, 1996). The period from planting to plant emergence is reduced along with an increase in soil temperature (Bizer, 1994). The application of nonwoven polypropylene cover enables earlier potato planting, forcing plant emergence and in the case of unfavourable thermal conditions, protects emerging plants against ground frosts. Presprouting seed-potatoes can be planted in the field when the soil warms up to 5-6°C. With the use of nonwoven polypropylene cover, planting can be started when soil temperature at the depth of 10 cm is about 3-4°C (Bizer, 1994; Lutomirska, 2006). Studies carried out in the Czech Republic, Poland, Bulgaria and Croatia showed that an increase in soil temperature as a result of applying nonwoven polypropylene cover in early crop potato culture shortened the period between planting to plant emergence by 2-8 days and forcing growth and development of plants in the later period (Table 1). Yearly variation of those differences are explained by environmental factors. The application of covering results in higher forcing of individual plant development phases in years with less favourable meteorological conditions during the initial period of potato growth. A higher increase of soil temperature under perforated polyethylene film, on average by 1-2°C as compared to nonwoven polypropylene, resulted in earlier occurrence of successive plant development phases by only 1-2 days (Ban et al. 2011; Cholakov & Nacheva, 2009; Hamouz et al., 2006; Lutomirska, 1995; Prośba-Białczyk & Mydlarski, 1998; Wadas & Kosterna, 2007a). Plants growing under nonwoven polypropylene were more uniform in size, higher, and developed a larger mass of aboveground parts as compared to cultivation without covering (Cholakov & Nacheva, 2009; Rekowska & Orłowski., 2000; Wadas & Kosterna 2007b; Wadas et al., 2009). Earlier plant development as a result of covering allow for the higher solar radiation interception and rapid enlargement of assimilation leaf area. A higher value of the leaf area index (LAI) has a favourable effect on the growth of tubers during the vegetation period, as well as on the final yield (Firman & Allen, 1989; Zrůst & Cepl, 1991; Zrůst et al. 1999). Forcing plant emergence as a result of applying covers results in more extensive ground cover by leaves only during early potato growth. Later, there were smaller differences as leaf senescence began under the cover sooner (Nelson & Jenkins, 1990). In the agro-meteorological conditions of east-central Poland, at the time of cover removal, the plants covered for two weeks after emergence were on average 9.4 cm higher, and after a 3-week

Country	Number of days	Reported by
Czech Republic	4-8	Hamouz et al., 2006
Poland	4-5	Lutomirska, 1995
	3-6	Prośba-Białczyk & Mydlarski, 1998; Wadas & Kosterna, 2007a
Bulgaria	2-7	Cholakov & Nacheva, 2009
Croatia	6	Ban et al., 2011

Table 1. Forcing of potato plant emergence as a result of nonwoven polypropylene covering

period of plant covering by 11.7 cm higher in comparison to plants cultivated without any covering, while the assimilation leaf area was 2 and 1.6 times higher, respectively. With the application of perforated polyethylene film, the plants were higher by 4.5 cm, on average, after 2 weeks from emergence, and by 7 cm after 3 weeks from emergence, in comparison to plants covered by nonwoven polypropylene fabric (Table 2). The type of cover applied had a smaller effect on the size of assimilation leaf area (Wadas & Kosterna 2007b; Wadas et al., 2009). The effect of covering on the assimilation leaf area depends to a high degree on meteorological conditions. In years with more favourable thermal conditions for early crop potato culture, leaving the covering (especially perforated film) for too long over plants after emergence can hinder the development of the assimilation leaf area, while in the lower temperature the effect is more favourable (Lutomirska & Szutkowska, 1999). When nonwoven polypropylene is removed too long after plant emergence, the transmision of photoactive radiation through this cover varied from 85 to 65%, depending on dust accumulation on the cover and water vapour condensation on the inner surface of the cover (Gimenez et al., 2002). A change of conditions in the initial period of plant growth as a result of applying nonwoven polypropylene cover could advance new potato harvest by up to 2-3 weeks, as compared to cultivation without plant covering (Bizer, 1994; Hamouz et al., 2005, as cited in Jaša, 1994; Sawicka & Pszczółkowski, 2002).

Specification		No covering	Nonwoven polypropylene	Perforated polyethylene film
Height of plants (cm)	2 weeks after plant emergence	17.7	27.1	31.6
	3 weeks after plant emergence	22.0	33.7	40.7
Weight of leaves per plant (g)	2 weeks after plant emergence	31	57	62
	3 weeks after plant emergence	52	77	90
Weight of stems per plant (g)	2 weeks after plant emergence	20	47	62
	3 weeks after plant emergence	33	71	109
Assimilatiom leaf area (cm²)	2 weeks after plant emergence	918	1867	1916
	3 weeks after plant emergence	1632	2665	2896
Leaf area index	2 weeks after plant emergence	0.49	0.99	1.02
	3 weeks after plant emergence	0.87	1.42	1.48

Table 2. Effect of nonwoven polypropylene and perforated polyethylene film covering on growth of early potato cultivars (Wadas & Kosterna, 2007b; Wadas et al., 2009)

4. Tuber yield

The use of nonwoven polypropylene cover makes possible to start harvesting early, to increase the yield of new potatoes and to reduce yield variability in successive years, in comparison to cultivation without plant covering. The production effect of applying cover reflected in an increase in the tuber yield depends, to a high degree, on soil and climatic conditions and on the potato harvesting date (Table 3).

Country	Increase in tuber yield (t ha⁻¹)	Reported by
Czech Republic	1.01-15.66	Hamouz et al., 2006, 2007
Poland	4.0-5.40	Prośba-Białczyk & Mydlarski, 1998
	2.87-9.97	Pszczółkowski & Sawicka, 1999
	6.36-10.04	Rekowska et al., 1999
	1.19-11.80	Wadas et al., 2001, 2008
Bulgaria	1.91-14.31	Cholakov & Nacheva, 2009
Croatia	1.85	Ban et al., 2011

Table 3. Effect of nonwoven polypropylene covering on marketable tuber yield of early potato

In conditions of east-central Poland (the Siedlce Region), covering the crop with nonwoven polypropylene resulted in an increase in marketable tuber yield after 60 days from planting very early cultivars of potato on average by 4.63 t ha⁻¹ (33%), while after 75 days from planting, the average yield increase was 3.72 t ha⁻¹ (13%) (Wadas et al., 2001). A similar increase in the early tuber yield of potato was obtained in the central part of Poland (Lutomirska, 1995). In later study carried out in east-central Poland, after 60 days from planting, the marketable tuber yield of early potato cultivars in cultivation with the use of nonwoven polypropylene was higher on average by 5.82 t ha⁻¹ (81%) (Wadas et al., 2008). In the Lublin Region, the average tuber yield increase as a result of nonwoven polypropylene covering amounted to 7.34 t ha⁻¹ (71.7%) after 60 days from planting, and 5.51 t ha⁻¹ (22.5%) at harvest date two weeks later (Pszczółkowski & Sawicka, 1999). In conditions of north-western Poland (the Szczecin Region), plant covering with nonwoven polypropylene resulted in increased marketable tuber yield after 55 days from planting on average by 9.84 t ha⁻¹ (52%), and after 65 days from planting - by 7.52 t ha⁻¹ (27%) (Rekowska et al., 1999). On the other hand, in south-western Poland (the Wrocław Region), where thermal conditions in spring are more favourable for production of early potatoes, plant covering with nonwoven polypropylene increased marketable tuber yield after about 60 days from planting very early potato cultivars on average by 4.5 t ha⁻¹ (74%), and when the harvest was carried out two weeks later, the difference in tuber yield amounted to 4.7 t ha⁻¹ (30%) (Prośba-Białczyk & Mydlarski, 1998). In the central part of the Czech Republic, plant covering with nonwoven polypropylene brought about an increase of marketable tuber yield on average by 6.28 t ha⁻¹ (50%) after 60 days from planting, and by 5.38 t ha⁻¹ (22%) when the harvest was carried out one week later (Hamouz et al., 2006, 2007), while in Croatia, the increase in tuber yield was 1.85 t ha⁻¹ (12%) (Ban et al., 2011). In Bulgaria, where the practice of growing cultures using nonwoven polypropylene covers is almost unknown, the increase in early potato yield as a result of using the cover amounted, on average, to 8.98 t ha⁻¹ (32%) after 60 days from plant emergence, and 6.21 t ha⁻¹ (16%) after 75 days from plant emergence (Cholakov & Nacheva, 2009). The study carried out in Poland, Germany and in eastern Canada (Quebéc) showed a higher favourable effect of covering with nonwoven polypropylene as compared to perforated polyethylene film in early crop potato culture. The application of nonwoven polypropylene made it possible to obtain a similar or even higher marketable tuber yield after 60 days from planting (Bizer, 1994; Demmler, 1998; Lutomirska, 1995; Michaud et al., 1990; Pszczółkowski & Sawicka, 2003; Rekowska & Orłowski 2000). Only in a year with a very cold spring, when the application of perforated polyethylene film obtained, in east-central Poland, a higher marketable tuber yield of early cultivars after 60 days from planting, on average by 4.18 t ha^{-1} (32%), while in years with warmer springs and at later harvesting date, yields obtained with application of perforated polyethylene film or nonwoven polypropylene were similar (Wadas et al. 2007). According to Bizer (1994), in years with dry springs, nonwoven polypropylene proves to be better covering, particularly on light soils, since a water shortage is likely to occur under perforated film. Cover use has the greatest effect on the potato tuber yield at very early harvest date, but when harvest is delayed, the effect of covering, reflecting in an increase in the new potato tuber yield in comparison to the cultivation without covering, is reduced. The study carried out in the central part of Czech Republic showed an increase in tuber yield as a result of applying covers in the period before the end of June, while along with the delay of the harvest date, the difference in the tuber yield in cultivation with and without covering was reduced, to reach an insignificant level by the end of June (Hamouz et al., 2004, 2005). The effect of applying nonwoven polypropylene cover in early crop potato culture depends to a high degree on meteorological conditions during the period of plant vegetation (Hamouz et al., 2004, 2006; Jabłońska-Ceglarek & Wadas, 2005; Lutomirska, 1995, 2006; Wadas et al., 2001, 2008). A higher increase of tuber yield as a result of applying the cover is obtained in years with less favourable thermal conditions in the initial period of potato growth. In east-central Poland in years with very cold springs, covering the crops with nonwoven polypropylene made it possible to obtain, after 60 days from planting, up to four times higher marketable tuber yield than in cultivation without covering (Jabłońska-Ceglarek & Wadas, 2005; Wadas et al., 2008). In the Czech Republic, potato cultivation under nonwoven polypropylene in the year with an exceptionally cold spring resulted in even six times higher marketable tuber yield after 60 days from planting (Dvořák et al., 2004; Hamouz et al., 2004, 2005). The application of covers assures to a high yield of potato tubers at an early harvest date, provided that the covers are removed at the proper time (Dvořák et al., 2007; Lutomirska & Szutkowska, 1999; Reust, 1980). Leaving the cover over the plants for too long after emergence can hinder development of assimilation leaf area and reduce the number of setting tubers. In conditions of east-central Poland, the length of plant covering period (two or three weeks after emergence) did not have any significant effect on the tuber yield after 60 days from planting, while for the harvest date delayed by two weeks, the yield was higher when the covers were removed two weeks after plant emergence (Wadas et al., 2008). The date of removing perforated polyethylene film had a greater effect on tuber yield than the date of removing nonwoven polypropylene, especially in years with warmer springs. The application of nonwoven polypropylene cover in the early crop potato culture assures not only a higher tuber yield, but also contributes to improvement of its structure by increasing the productivity of marketable tuber fractions, and simultaneously increasing share of large tubers in the yield, with diameters above 50 mm (Ban et al., 2011; Prośba-Białczyk & Mydlarski, 1998; Pszczółkowski & Sawicka, 2003; Rekowska et al., 1999; Wadas et al., 2001, 2008).

5. Tuber quality

A change in conditions for the initial growth and development of potato plants by applying nonwoven polypropylene or perforated polyethylene film cover affects not only the volume of the tuber yield, but also the chemical composition of tubers. Forcing of plant growth as a result of using covers results in an increase in dry matter and starch content in tubers. While applying nonwoven polypropylene, the dry matter content in tubers was higher by 0.69-2.17%, and starch by 0.45-1.46% than in cultivation without plant covering (Dvořák et al., 2006, 2008; Hamouz et al., 2006; Jabłońska-Ceglarek & Wadas, 2005; Wadas et al., 2003, 2004, 2006). A greater beneficial effect of applying covers in the form of an increase of dry matter content in tubers is found for early harvest date and in years with high air temperatures and abundant sunshine. According to other authors, the application of nonwoven polypropylene or perforated polyethylene film covers created less favourable conditions for accumulation of dry matter in tubers of very early potato cultivars (Prośba-Białczyk & Mydlarski, 1998; Sawicka & Pszczółkowski, 2005). Changed conditions for initial growth and development of potato plants as a result of applying the covers resulted in an increase in content of total sugars, reducing sugars and saccharose in tubers (Sawicka & Pszczółkowski, 2005). The study did not show any effect of applying nonwoven polypropylene in early crop potato culture on protein accumulation in tubers. On the other hand, a tendency was observed towards an increase in the ascorbic acid (vitamin C) concentration and a decrease in the concentration of carotenoids and polyphenols (Dvořák et al., 2006, 2008; Jabłońska-Ceglarek & Wadas, 2005; Lachman et al., 2003; Wadas et al., 2003, 2004, 2006). This method of potato cultivation resulted in a lower content of ascorbic acid in tubers as compared to cultivation without plant covering (Prośba-Białczyk & Mydlarski, 1998). Plant growth forcing as a result of nonwoven polypropylene covering contributes to an improvement of the tuber quality by reducing the concentration of nitrates, aspecially for the early harvesting date. In this method of potato cultivation, the content of nitrates was lower by 29-239 mg NO_3 in 1 kg of fresh weight of tubers (Dvořák et al., 2006; Lachman et al., 2003; Wadas et al., 2005). Covering the plants with nonwoven polypropylene also created very favourable conditions for accumulation of phosphorus and potassium in tubers (Wadas et al., 2007, 2008). The application of covers in early crop potato culture should be also considered in the aspect of plant health. Higher soil moisture under the cover can contribute to an increased occurrence of tubers infected with *Streptomyces scabies* and *Rhizoctonia solani* and to a faster rate of *Phytophthora infestans* spreading (Pszczółkowski & Sawicka, 1998).

6. Cost effectiveness of the production

The application of nonwoven polypropylene cover in potato production requires higher inputs incurred related not only to the purchase of nonwoven polypropylene, but also to labour input for its spreading and removing, as well as for harvesting the crops (Prośba-Białczyk et al., 1997; Pszczółkowski et al., 2000/2001; Wadas, 2003; Wadas et al., 2003, 2006; Wadas & Sawicki, 2009). Increasing production inputs is effective when the value of the tuber yield increase obtained as a result of plant covering is higher than the costs incurred. German studies showed that for the cost of purchasing nonwoven polypropylene amounting to 2000 DM per 1 ha, when the cover is used 2-2.5 times, yearly costs ranged from 800 and 1000 DM per 1 ha, while for the cost of purchasing perforated polyethylene film amounting to 1400 DM per 1 ha, and using the cover 1.5-2 times, yearly costs ranged from 700 to 1050 DM. Therefore, yearly costs of using both covers were similar (Demmler, 1998). The amount of actually incurred costs depends, first of all, on how many times the nonwoven polypropylene cover is reused. The reuse rate of nonwoven polypropylene depends on its mechanical damage, sun radiation and degree of contamination. In the Czech Republic, at the cost of purchasing nonwoven polypropylene amounting to 1000-1200 EURO per 1 ha, its two-time application proved cost-effective for farmers (Hamouz & Dvořák 2004). In Poland, the cost of purchasing nonwoven polypropylene calculated for three seasons of its use amounted to PLN 5200-6300 per 1 ha (Prośba-Białczyk et al., 1997; Wadas, 2003). In south-western Poland, the production cost of early potatoes under nonwoven polypropylene was higher, depending on the year, between 85-92% as compared to cultivation without the cover, and the cost of nonwoven polypropylene accounted for 37-41% of incurred costs (Prośba-Białczyk et al., 1997). Such method of producing early potatoes in east-central Poland required incurring costs which were higher by 47-89%, and the cost of purchasing the nonwoven polypropylene accounted for 24-40% of direct costs. Due to a higher price, the direct costs of producing early potatoes under nonwoven polypropylene were higher by 18-25% than under perforated polyethylene film (Wadas, 2003; Wadas et al., 2003, 2006; Wadas & Sawicki, 2005, 2009). While assessing the cost-effectiveness of early potato production under nonwoven polypropylene cover, it is not only the sum of incurred costs which is important, but also unit costs, which provide information about the level of selling price which will balance the costs incurred. In south-western Poland, the costs of producing 1 kg of tubers under nonwoven polypropylene were 1.3-1.5 times higher, and in a year with very favourable conditions for early crop potato culture, it was almost the same as in the cultivation without covering. In the agro-meteorological conditions of east-central Poland, covering the crop with nonwoven polypropylene increased unit costs of production by 1.2 to 2.1 times. Unit costs of production were lower than in cultivation without covering only in the year with unfavourable thermal conditions for early crop potato culture, due to high yields obtained in cultivation under nonwoven polypropylene. Unit costs of production under nonwoven polypropylene were 1.3 to 1.6 times higher compared with perforated polyethylene film. Only in one year which was very favourable for early crop potato culture were they almost the same to that of applying perforated polyethylene film. It is more efficient to increase imputs for early potato production by applying nonwoven polypropylene in less favourable thermal conditions during the initial period of potato growth. In such a case, a significant yield increase in cultivation under covers balances the costs incurred and makes it possible to obtain a higher direct surplus from production than without covering (Table 4).

The cost-effectiveness of early potatoes production under nonwoven polypropylene cover depends on the income-to-costs ratio. Applying the nonwoven polypropylene in the early crop potato culture ensures high cost-effectiveness of production in years with cold springs. In conditions favouring rapid growth of potatoes, production costs of 1 kg tubers under cover are higher, which makes production less profitable in comparison to cultivation without plant covering.

Specification	Nonwoven polypropylene			Perforated polyethylene film		
	Years with warm spring		Year with cold spring	Years with warm spring		Year with cold spring
	2002	2003	2004	2002	2003	2004
Increase in production costs (PLN ha^{-1})	4067.5	4032.2	4330.0	2229.1	2340.8	2665.3
Increase in tuber yield (t ha^{-1})	6.09	1.79	9.59	2.79	2.39	13.78
Value of additional tuber yield (PLN ha^{-1})	10953.0	3585.0	16311.5	5022.0	4780.0	23413.2
Marginal effectiveness	2.67	0.89	3.76	2.22	2.03	8.77

Table 4. Cost effectiveness of nonwoven polypropylene and perforated polyethylene film covers use in early potato production (Wadas & Sawicki, 2009)

7. Conclusions

Success in early crop potato culture depends to a high degrre on the soil and air temperature in the initial period of plant growth. Obtaining early yields from field production is possible in soil and climatic conditions which enable early planting and rapid plant growth. The unfavourable effect of low temperatures in the initial period of potato growth can be reduced by the application of nonwoven polypropylene cover directly on the planted field. With the use of covers, potato planting can be started when the soil temperature at the depth of 10 cm is about 3-4°C. The application of covers enables earlier potato planting, forcing plant emergence and the growth and development of plants in the later period, and consequently, results in earlier setting of tubers, rapid yield gain and reduction of the yield variability in any years. A change in conditions during the initial period of plant growth as a result of applying covers could advance new potato harvest by up to 2-3 weeks. Such a method of production requires higher input incurred, while the effect of applying covers, reflected in an increase in the tuber yield, to a high degree depends on soil and climatic conditions. An increase in production inputs by the application of nonwoven polypropylene cover is more effective in less favourable thermal conditions in the initial period of potato growth. A considerable tuber yield increase in cultivation under cover results in such a case in decrease of unit costs and consequently, the cost-effectiveness of production is higher than without covering. In conditions favouring rapid potato growth, the application of covers increases unit costs, which makes production less profitable as compared to cultivation without covering. The application of nonwoven polypropylene cover facilitates a significant increase in income from potato production at a very early harvest date. Along with a delay in harvesting, the effect of applying the cover, reflected in an increase in the tuber yield, decreases in comparison to cultivation without covering.

8. Acknowledgment

Krystyna Struk, Director of Agricultural Experimental Station in Zawady, is acknowledged for her valuable technical assistance during the field experiments and during harvest operations. The author would like to acknowledge the Reviewers and Editors for their valuable comments and suggestions for this paper.

9. References

Ban, D.; Vrtačić, M.; Goreta Ban, S.; Dumičić G.; Oplanić, M.; Horvat, J. & Žnidarčić. D. (2011). Effect of variety, direct covering and date of harvest on the early potato growth and yield. *Proceedings of 46th Croatian and 6th International Symposium on Agriculture*, pp. 496-500, ISBN 978-953-6135-71-3, Opatija, Croatia, February 14-18, 2011

Bizer, E. (1994). Frühkartoffelanbau unter Vlies und Folie. *Kartoffelbau*, Vol. 45, No. 2, pp. 462-466

Bizer, E. (1997). Ernteverfruhung durch Vliesabdeckung. *Kartoffelbau*, Vol. 48, No. 1/2, pp. 60-61

Cholakov, T.L. & Nacheva, E.K. (2009). Results from using polypropylene cover in production of early potatoes. *Acta Horticulturae*, Vol. 839, pp. 603-608, ISSN 0567-7572, ISBN 978-90-66050157-7

Demmler, D. (1998). Vergleich von Folie und Vlies zur Ernteverfrühung in Frühkartoffeln. *Kartoffelbau*, Vol. 49, No. 12, pp. 429-430, ISSN 0022-9156

Dvořák, P.; Hamouz, K.; Bicanova, E. & Prasilova, M. (2007). Effect of the date of polypropylene textile removal and site on yield-forming components of early potatoes. *Scentia Agriculturae Bohemica*, Vol. 38, No. 4, pp. 162-167, ISSN 1211-3174

Dvořák, P.; Hamouz. K.; Čepl, J. & Pivec, J. (2004). The non-woven fleece as an implement for acceleration of early potatoes harvest. *Scientia Agriculturae Bohemica*, Vol. 35, No. 4, pp. 127-130, ISSN 1211-3174

Dvořák, P.; Hamouz, K.; Jůzl, M. & Erhartowa, D. (2006). Influence of row covering with non-woven textile on tubers quality in early potatoes. *Zeszyty Problemowe Postępów Nauk Rolniczych*, No. 511, pp. 225-231, ISSN 0084-5477

Dvořák, P.; Hamouz, K. & Lachman, J. (2008). Effect the polypropylene textile cover on tubers quality of early potatoes. *Proceedings of 43rd Croatian and 3ed International Symposium on Agriculture*, pp. 628-631, ISBN 978-953-6135-68-4, Opatija, Croatia, February 18-21, 2008

Firman, D. M. & Allen, E. J. (1989). Relationship between light interception, ground cover and leaf area index in potatoes. *The Journal of Agricultural Science*, Vol. 113, pp. 355-359, ISSN 0021-8596, EISSN 1469-5146

Friessleben, R. (1984). Untersuchungen zum Anbau von Speisefrühkartoffeln inter perforierter Polyäthylenfolie. *Archiv für Acker- und Pflanzenbau und Bodenkunde*, Vol. 28, pp. 133-142, ISSN 0365-0340

Gimenez, C.; Otto, R.F. & Castilla, N. (2002). Productivity of leaf and root vegetable crop under direct cover. *Scientia Horticulturae*, Vol. 94, No. 1/2, pp. 1-11, ISSN 0304-4238

Gordon, R.; Brown, D.M. & Dixon, M.A. (1997). Estimating potato leaf area index for specific cultivars. *Potato Research*, Vol. 40, pp. 251-266, ISSN 0014-3065

Hamouz, K. & Dvořák, P. (2004). Influence of white fleece on the yield formation of early potatoes. *Proceedings of 39th Croatian Symposium on Agriculture*, pp. 395-396, Croatia, February 17-20,2004

Hamouz, K.; Dvořák, P.; Čepl, J. & Pivec, J. (2005). The effect of polypropylene fleece covering on the yield of early potatoes. *Horticultural Science (Prague)*, Vol. 32, No. 2, pp. 56-59, ISSN 0862-867X

Hamouz, K.; Dvořák, P. & Erhartova, D. (2007). Effect of polypropylene covering on the yield formation dynamics of early potatoes. *Acta Phytotechnica et Zootechnica*, Vol. 3, pp. 57-60, ISSN 1336-9245

Hamouz, K.; Dvořák, P. & Šařec, O. (2004). Efficiency of white fleece during the cultivation of early potatoes. *Zeszyty Problemowe Postępów Nauk Rolniczych*, Vol. 500, pp. 271-276, ISSN 0084-5477

Hamouz, K.; Lachman, J.; Dvořák, P. & Trnková, E. (2006). Influence of non-woven fleece on the yield formation of early potatoes. *Plant Soil and Environment*, Vol. 52, No. 7, pp. 289-294, ISSN 1214-1178

Hamouz, K. & Rybáček V. (1988). Porous foil mulching of the stands of very early potatoes. *Rostlinná Výroba*, Vol. 34, pp. 1095-1102, ISSN 0035-8371

Jabłońska-Ceglarek, R. & Wadas, W. (2005). Effect of nonwoven polypropylene covers on early tuber yield of potato crops. *Plant Soil and Environment*, Vol. 51, No. 5, pp. 226-231, ISSN 1214-1178

Jenkins, P.D. & Gillison, T.C. (1995). Effects of plastic film covers on dry-matter production and early tuber yield in potato crop. *Annals of Applied Biology*, Vol. 127, pp. 201-213, ISSN 0003-4746

Kubiak, K. & Gaziński, B. (1996). The early potato marked in Poland. *Postępy Nauk Rolniczych*, No. 5, pp. 43-51, PL ISSN 0032-5547

Lachman, J.; Hamouz, K.; Hejtmánková, A.; Dudjak, J.; Orsák, M. & Pivec, V. (2003). Effect of white fleece on the selected quality parameters of early potato (*Solanum tuberosum*, L.) tubers. *Plant Soild and Environment*, Vol. 49, No. 8, pp. 370-377, ISSN 1214-1178

Lang, H. (1984). Folieneinsatz im Frühkartoffelbau für Sicherheit, Ertrag und Qualität. *Kartoffelbau*, Vol. 35, pp. 65-69, ISSN 0022-9156

Lutomirska, B. (1995). Usage of agrotextile to forceing the yielding of potatoes. *Ziemniak Polski*, No. 3, pp. 13-19, ISSN 1425-4263

Lutomirska, B. (1995). Acceleration of the accumulation of the commercial yield in early cultivars throught the use of covers. *Proceedings of 28th Scientific Conference Potato cultivation technology and selected problems from storage*, pp. 49-54, Research Institute of Potato, Bonin, March 9-10, 1995

Lutomirska, B. (2006). Forcing of very early potatoes harvest. *Ziemniak Polski*, No. 1, pp. 12-15, ISSN 1425-4263

Lutomirska, B. & Szutkowska, M. (1999). Assimilation area and early yield under the cover application in the potato cultivation. *Proceedings of Conference Potato for consumption and for food processing – agricultural and storage factors ensuring quality*, pp. 169-171, Plant Breeding and Acclimatization Institute, Radzików, February 23-25, 1999

Michaud, M. H.; Dubé, P. A. & Bégin, S. (1990). Influence of floating row covers on microclimate for production of early potatoes (*Solanum tuberosum* L.). *American Potato Journal*, Vol. 67, pp. 565-566, ISSN 0003-0589

Nelson, D. G. & Jenkins, P. D. (1990). Effects of physiological age and floating plastic film on tuber dry-matter percentage of potatoes, cv. Record. *Potato Research*, Vol. 33, pp. 159-169, ISSN 0014-3065

Nishibe, S.; Satoh, M.; Mori, M.; Isoda, A. & Nakaseko, K. (1989). Effect of climatic conditions on intercepted radiation and some growth parameters in potato. *Japanese Journal of Crop Science*, Vol. 52, No. 2, pp. 171-179, ISSN 0011-1848

Prośba-Białczyk, U. & Mydlarski, M. (1998). Growth potato on early harvest under cover with polypropylene sheets. *Fragmenta Agronomica*, No. 1 (57), pp. 74-84, PL ISSN 0860-4088

Prośba-Białczyk, U.; Paluch, F. & Mydlarski, M. (1997). Economic effectiveness of early potato production under woven polypropylene sheets. *Bibliotheka Fragmenta Agronomica*, No. 3, pp. 181-188, ISSN 0860-4088

Pszczółkowski, P.; Harasim, A. & Sawicka, B. (2000/2001). Economic performance of the production technologies for early consumption potatoes harvested on different dates. *Roczniki Nauk Rolniczych*, Ser. G, Vol. 89, No. 1, pp. 89-99, ISSN 0080-3715

Pszczółkowski, P. & Sawicka, B. (1998). Application of shields and various techniques in cultivation of early potato varieties in a bearing of plant health. *Roczniki AR Poznań*, CCCVII, Rolnictwo, No. 52, pp. 191-196, ISSN 0137-1754

Pszczółkowski, P. & Sawicka, B. (1999). Tuber yield of very early potato cultivars cultivated under polypropylene fibre cover. *Proceedings of Science Conference Cultivation of horticultural plants at the threshold of 21st century*, pp. 31-34, AR Lublin, February 4-5, 1999

Pszczółkowski, P. & Sawicka, B. (2003). Productivity of early potato cultivars cultivated under coverage. Part. I. Yield and its structure. *Acta Scientiarum Polonorum*, *Agricultura*, Vol. 2, No. 2, pp. 61-72, ISSN 1644-0625

Rekowska, E. & Orłowski, M. (2000). Effect of cultivation methods on the quantity and quality of the yield of early potato. *Annales UMCS*, Sectio EEE, vol. VIII, Suppl.: pp. 129-135, ISSN 1233-2127

Rekowska, E.; Orłowski, M. & Słodkowski, P. (1999). Yielding of early potato as affected by covering application and the terms of harvest. *Zeszyty Problemowe Postępów Nauk Rolniczych*, Vol. 466, pp. 181-189, ISSN 0084-5477

Reust, W. (1980). Culture de pommes de terre primeur sous film en matière plastique. *Revue Suisse Agriculture.*, Vol. 12, No. 2, pp. 61-64 , ISSN 0375-1325

Sale, P.J.M. (1979). Growth of potatoes (*Solanum tuberosum* L.) to small tuber stage as related to soil temperature. *Australian Journal of Agricultural Research*, Vol. 30, No. 4, pp. 667-675, ISSN 0004-9409

Sawicka, B. & Pszczółkowski, P. (2002). Progress in under cover technology of early potato cultivars production. *Pamiętnik Puławski*, Vol. 130, No. 2, pp. 673-683, ISSN 0552-9778

Sawicka, B. & Pszczółkowski, P. (2005). Dry matter and carbohydrates content in the tubers of very early potato varieties cultivated under coverage. *Acta Scientiarum Polonorum, Hortorum Cultus*, Vol. 4, No. 2, pp. 111-122, ISSN 1644-0692

Wadas, W. (2003). The economic effectiveness of early potato production under agrotextile covers. *Pamiętnik Puławski*, Vol. 133, pp. 207-214, ISSN 0552-9778

Wadas, W.; Jabłońska-Ceglarek, R. & Kosterna, E. (2003). The effect of nonwoven polypropylene covering in cultivation of very early potato cultivars on the content of some nutrient components in immature tubers. *Żywność (Nauka, Technologia, Jakość)*, No. 3 (36), pp. 110-118, ISSN 1425-6959

Wadas, W.; Jabłońska-Ceglarek, R. & Kosterna E. (2004). Effect of plastic covering and nitrogen fertilization on yield and quality of early potatoes. *Folia Horticulturae*, Vol. 16, No. 2, pp. 41-48, PL ISSN 0867-1761

Wadas, W.; Jabłońska-Ceglarek, R. & Kosterna E. (2005). The nitrates content in early potato tubers depending on growing conditions. *Electronic Journal of Polish Agricultural Universities*, Horticulture, Vol. 8, Issue 1, ISSN 1505-0297, Available online: http://www.ejpau.media.pl/volume8/issue1/art-26.html

Wadas, W.; Jabłońska-Ceglarek, R.; Kosterna, E. & Łęczycka, T. (2007). The potassium content in early potato tubers depending on cultivation method. *Roczniki Akademii Rolniczej w Poznaniu, Ogrodnictwo*, Vol. 41, pp. 643-647, PL ISSN 0137-1738

Wadas, W.; Jabłońska-Ceglarek, R. & Kurowska, A. (2008). Effect of using covers in early crop potato culture on the content of phosphorus and magnesium in tubers. *Journal of Elementology*, Vol. 13, No. 2, pp. 275-280, ISSN 1664-2296

Wadas, W.; Jabłońska-Ceglarek, R. & Rosa, R. (2001). A posssibility of increasing the yield of young potato tubers by using a polypropylene fibre covers. *Electronic Journal of Polish Agricultural Universities*, Horticulture, Vol. 4, Issue 2, ISSN 1505-0297, Available online: http://www.ejpau.media.pl/series/volume4/issue2/horticulture/art-06.html

Wadas, W. & Kosterna, E. (2007a). Effect of perforated foil and polypropylene fibre covers on development of early potato cultivars. *Plant, Soil and Environment*,Vol. 53, No. 3, pp. 136-141, ISSN 1214-1178

Wadas, W. & Kosterna, E. (2007b). Effect of perforated foil and polypropylene fibre covers on assimilation leaf area of early potato cultivars. *Plant, Soil and Environment*,Vol. 53, No. 7, pp. 299-305, ISSN 1214-1178

Wadas, W.; Kosterna, E. & Kurowska, A. (2009). Effect of perforated foil and polypropylene fibre covers on growth of early potato cultivars. *Plant, Soil and Environment*,Vol. 55, No. 1, pp. 33-41, ISSN 1214-1178

Wadas, W.; Kosterna, E. & Sawicki, M. (2007). Productivity of early potato cultivar in the cultivation under perforated foil and polypropylene fibre covers. *Fragmenta Agronomica*, No. 2, pp. 364-372, PL ISSN 0860-4088

Wadas, W.; Kosterna, E. & Sawicki, M. (2008). Effect of perforated film and polypropylene nonwoven covering on the marketable value of early potato yield. *Vegetable Crops Research Bulletin*, Vol. 69, pp.51-61, PL ISSN 1506-9427

Wadas, W.; Kosterna, E. & Żebrowska, T. (2006). The effect of using covers in cultivation of early potato cultivars on the content of some nutrients in tubers. *Zeszyty Problemowe Postępów Nauk Rolniczych*, No. 511, pp. 233-243, ISSN 0084-5477

Wadas, W. & Sawicki, M. (2005). Estimation of early potato production profitability in mid-eastern Poland conditions. *Pamiętnik Puławski*, Vol. 139, pp. 289-297, ISSN 0552-9778

Wadas, W. & Sawicki, M. (2009). The economic effectiveness of early potato production depending on the kind of cover. *Polish Journal of Agronomy*, Vol. 1, pp. 56-61, ISSN 2081-2787

Wadas, W.; Sawicki, M. & Kosterna, E. (2006). Production costs of the potatoes for early harvest under perforated foil and polypropylene fibre covers. *Zeszyty Problemowe Postępów Nauk Rolniczych*, Vol. 511, pp. 417-427, ISSN 0084-5477

Zrůst, J. & Cepl, J. (1991). Dependence of early potato yield on some growth characteristics. *Rostlinná Výroba*, Vol. 37, No. 11, pp. 925-933, ISSN 0035-8371

Zrůst, J.; Hlušek, J.; Jůzl, M. & Přichystalowá, V. (1999). Relationship between some chosen growth characteristics and yield of very early potato varieties. *Rostlinná Výroba*, Vol. 45, No, 11, pp. 503-509, ISSN 0035-8371

10

Influence of Low Pressure on the Stability of Polypropylene Electrets Films

Asya Viraneva, Temenuzhka Yovcheva and Georgi Mekishev
University of Plovdiv, Department of Experimental Physics
Bulgaria

1. Introduction

Electret is an important cross-scientific subject of dielectric physics, material science, sensor engineering, medical and bio-engineering (Sessler G., Sessler et al., Nalwa H., Goel M.). Over the years, considerable interest has been shown in the surface potential decay of corona-charged polymeric materials. Besides the electrets material and conditions of producing electrets the surface potential decay depends on number of factors under which the electrets have been stored or used, for example temperature, humidity, pressure etc. (Zhongfu et al., Gang-Jin et al., Ribeiro et al.).

The pressure effect has been investigated in number of papers. But there are only a few papers dealing with the influence of pressure lower than atmospheric on the electrets behaviour (Shepard et al., Wild et al., Gubkin et al., Draughn et al., Palaia et al., Mekishev et al., 2003, Mekishev et al., 2005). Moreover, the conclusions which have been made are contradictory. Some of the authors assume that charge decay is a result of sparking breakdown in the air according to Paschen's low and the others assume ions desorption from electret surface. Sheppard and Stranathan (Shepard et al.) investigated the effect of ambient pressure on the electret charge of thermoelectrets and showed that in the region of a few torr to 200 torr the charge of electrets increases proportionally to pressure. The linear dependence of the surface charge on pressure is in keeping with Henry's law of adsorption (Adamson et al.). Analogous linear dependence of electret surface potential on pressure in the range of 20 torr to 200 torr is observed in (Mekishev et al., 2003).

Wild and Stranathan (Wild et al.) studied the influence of external treatments on carnauba wax electret behaviour including the pressure effect in a wide range of pressures ranging from 10⁻⁵ torr to 3000 torr. Except for the plateaus at both low and high pressures, the curve obtained is similar to the curve of the sparking potential of air versus the logarithm of air pressure. Draughn and Catlin (Draughn et al.) investigated the effect of low pressure on surface charge of polystyrene and mylar electrets. All samples were stored with their surfaces exposed to the environment, i.e. unshielded. Authors conclude that the decrease in effective surface charge observed when electrets are exposed to low pressure is not the result of a spark breakdown between the electret surface and nearby conductors. The data indicate that the charge drops are due to ion desorption. After a discussion by B. Gross (Gross B.), Palaia and Catlin (Palaia et al.) investigated again the behaviour of oriented polystyrene film and mylar electrets at low pressure. The results were found to be as predicted by the spark breakdown theory. It should be noted that in the papers referred above the method of dissectible capacitor was used to measure the effective surface charge. In (Mekishev et al., 2003, Mekishev et al., 2005) the electrets surface potential was measured by the vibrating-electrode method with compensation. The measurement system was located in a chamber connected to a vacuum pump and a vacuum gauge to measure the pressure in the chamber. After charging, the samples were placed in the measurement system and the initial electrets surface potential was measured. Then the pump produced low pressure, and the electrets surface potential vs storage time relationship at various pressure (20, 50, 100, 200 and 300 Torr) was measured for 1 h. It was shown (Mekishev et al., 2003) that the dependence of the surface potential on pressure is described by an equation which is analogous to the Langmuir law of adsorption. Furthermore, it was found in (Mekishev et al., 2005) that the time dependence of the electret surface potential at various pressures is described well by the differential equation for desorption. The authors proposed that ions desorption from the electret surface is most likely to occur. Polypropylene (PP) is one of the polymers, which has been widely used for preparing electrets, because of its excellent dielectric properties (Ono et al., Mohmeyer et al.).

The present paper reports the results of investigations on the low pressure effect on the behaviour of corona charged PP electrets. The PP films are charged in a positive or a negative corona at different charging temperatures: T=20°C and T=90°C. In the vacuum chamber some electrets were stored between two plate shorted electrodes with various air gaps and the other ones were put only onto a metal electrode. The time and temperature dependences of the surface potential of electrets were also studied.

2. Experimental

2.1 Used material and sample preparation

In the present paper isotactic PP films with thickness of 20μm produced by "Assenova Krepost" LTD – Bulgaria were investigated. Initially, the PP films were cleaned in an ultrasonic bath with alcohol for 4 minutes then washed in distilled water and dried on filter paper under room conditions. Samples of 30mm diameter were cut from the films. All the samples were put onto the same diameter metal pads.

2.2 Corona charging and surface potential measurement of the samples

The charging of the samples in a corona discharge was carried out by means of a conventional corona triode system (Jeda et al.) consisting of a corona electrode (needle), a grounded plate electrode and a grid placed between them – Fig. 1. The distance between the corona electrode and the grid was 10mm and the grid to grounded plate electrode distance was 3mm.

The samples 1 together with their metal pads were placed on the grounded plate electrode 4 and were charged for 1 minute at different charging temperatures (T=20°C and T=90°C). Positive or negative 5kV voltage was applied to the corona electrode 2. Voltages of the same polarity as that of the corona electrode from a power supply NB-825 were applied to the grid 3 and their values were 350V, 500V, 650V, 800V and 950V respectively. Electrets surface potential was measured by the method of the vibrating electrode with compensation (Reedyk et al.) by which the estimated error was better than 5%.

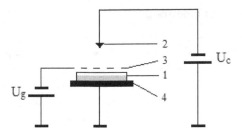

Fig. 1. Arrangement of the corona charging set-up.
1 – sample on a metal pad; 2 – corona electrode; 3 – grid; 4 – grounded plate electrode;
U_g – grid voltage power supply; U_c –corona voltage power supply.

2.3 Low pressure treatment

After measuring the initial surface potential V_0, the samples, together with their metal pads, were placed for 30 minutes in a vacuum chamber under various low pressures.

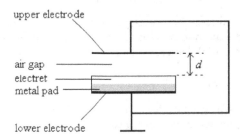

Fig. 2. Schematic diagram of the electret storage in the vacuum chamber between two short circuited plate electrodes (d is the air gap thickness).

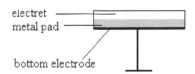

Fig. 3. Schematic diagram of the electret storage in the vacuum chamber in the absence of a second electrode.

The pressures in the vacuum chamber were 0.1mbar, 1mbar, 10mbar, 20mbar, 66mbar, 132mbar and 1000mbar. In the vacuum chamber the electrets were placed between two short circuited plate electrodes (Fig. 2) or deposited on metal pads in the absence of a second electrode (Fig. 3). After removing the electrets from the vacuum chamber, their surface potentials V were measured again and the normalized surface potentials V / V_0 were calculated.

3. Results and discussion

3.1 Results

3.1.1 Influence of the type of electrode

Two groups of experiments were carried out. The first group samples were non-metalized and placed on metal pads and the ones from the second group – metalized on one side. All samples were corona charged to the initial value of about 660V for the surface potential. The samples were then placed in a vacuum chamber at pressure of 0.1mbar for 30 minutes. The results obtained are presented in Table 1, where V_0 is the initial surface potential and V is the surface potential measured just after the samples were taken out of the vacuum chamber.

Samples	Surface potential		V / V_0
	Just after charging, V_0, V	After storage at 0.1mbar for 30 min., V, V	
Non-metalized	655±7	186±6	0.28±0.01
One-side metalized	671±7	189±7	0.28±0.01

Table 1. Surface potential for PP samples stored at pressure of 0.1mbar.

The results displayed in Table 1 show that the one-side metalized samples do not behave differently, compared with the non-metalized and placed on metal pads ones. In both cases the normalized surface potential decreases to about 0.28.

3.1.2 Charging time influence

The influence of the charging time on the surface potential of the electrets was studied. For that purpose the PP samples were corona charged using a three-electrode system (Fig. 1) at different charge time t_{ch} – 1, 5, 10, 20 and 30 minutes. The voltage U_c = -5kV was applied to the corona electrode and U_g = -650V to the grid.

Charging time, min.	V_0, V
1	657±7
5	664±6
10	660±6
20	650±5
30	648±9

Table 2. Surface potential of PP samples at different values for the charging time.

After the charging the initial surface potential V_0 of the samples was measured by the vibrating electrode method with compensation (Reedyk et al.). The measured surface potential values are presented in Table 2. The results obtained show that the charging time to 30 minutes does not influence the surface potential of the PP films.

3.1.3 Influence of the storage time under low pressure

The influence of sample storage time in the vacuum chamber on the surface potential decay was studied. For this purpose the samples were corona charged (Fig. 1) at a corona voltage of U_c = -5kV and a grid voltage of U_g = -650V. The initial surface potential V_0 was measured by the vibrating electrode method with compensation. Then the samples were placed in a vacuum chamber at a pressure of 0.1mbar at different time periods - 1, 15, 30 and 60 minutes. After the samples being taken out of the vacuum chamber their surface potential V was measured again. The results obtained are presented in Table 3.

Storage time at 0.1mbar, min.	Surface potential		V / V_0
	V_0, V	V, V	
1	642±9	187±21	0.29±0.04
15	639±9	184±18	0.29±0.03
30	648±5	172±5	0.27±0.01
60	646±4	177±20	0.27±0.03

Table 3. Surface potential of PP samples at different values of storage time, at 0.1 mbar.

It can be seen from the results in the table that the surface potential decreases within the first minute and that further storage at the reduced pressure does not bring any change. Based on the results obtained in parts 3.1.1, 3.1.2, 3.1.3 the next experiments were carried out with non-metalized samples being charged for a minute in a positive or a negative corona and being stored in the vacuum chamber for 30 minutes (for greater security).

The published papers discuss two main processes causing electret discharge at low pressure - sparking breakdown in air according to Paschen's law and ions desorption from electret surface. In order these processes to be studied two groups of experiments with different samples bonding were carried out. In the vacuum chamber the electrets were placed between two short circuited plate electrodes (Fig. 2) or deposited on metal pads in the absence of a second electrode (Fig. 3). The time and temperature dependences for the surface potential of polypropylene electrets charged in a negative corona at room temperature and temperature of 90°C were also studied.

3.1.4 Electrets obtained in a corona charging and stored at low pressures between two short-circuited plate electrodes with various air gaps between the electrodes

All the samples were divided in five groups according to the thickness of the air gap. Each group was split into five series depending on the value of the initial surface potential.

The samples prepared were corona charged at room temperature by the three-electrode system (Fig. 1). Negative 5kV voltage U_c was applied to the corona electrode and to the grid – negative voltage U_g of 350V, 500V, 650V, 800V and 950V respectively. The electrets were thereby charged to different values of the initial surface potential V_0. After the samples were corona charged their initial surface potential was measured. Then the samples together with their metal pads were placed between two short-circuited flat electrodes, at a different thickness of the air gap (d – 0.10mm, 0.28mm, 0.84mm, 1.69mm and 3.00mm). The system obtained in that way was stored in a vacuum chamber for 30 minutes at various low pressures - 0.1mbar, 1mbar, 10mbar, 20mbar, 66mbar, 132mbar and 1000mbar. For each of these pressures, after taking the samples out of the vacuum chamber their surface potential V was measured again. Normalized surface potential V/V_0 vs normalized pressure p/p_0 for negatively charged PP samples stored at two air gaps - d =0.10mm and d =3.00mm respectively - are presented in Fig. 4 and Fig. 5.

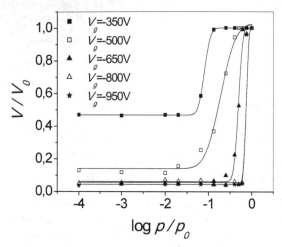

Fig. 4. Dependence of the normalized surface potential on normalized pressure for negatively charged PP samples. The air gap is d =0.10mm.

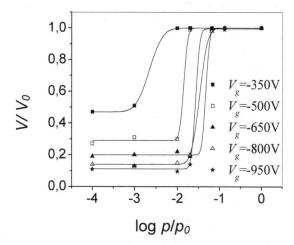

Fig. 5. Dependence of the normalized surface potential on normalized pressure for negatively charged PP samples. The air gap is d = 3.00mm.

Each figure point corresponds to an average value obtained by the measurement of 6 samples. The maximum deviation from the average values determined at confidence level

95% is 5%. The V_0 symbol stands for the initial value of the surface potential measured just after the samples charging and p_0 is the atmospheric pressure.

Analogous dependences were obtained for the other air gap values.

The results presented in Fig. 4 and Fig. 5 show the following features:

- For each curve three parts are observed. At high pressures the normalized surface potential stays constant and close to one (i.e. the surface potential at high pressures is approximately equal to the initial surface potential); at low pressures V / V_0 is also constant, however, with much lower values. For all the values of the initial surface potential, i.e. for all the series of samples there is a range of pressures within which the surface potential V_0 sharply decreases.
- The higher the initial surface potential, the higher the pressure at which the sharp decrease takes place, i.e. as the initial surface potential grows the curves are displaced to the right.
- With the increase of the air gap the decrease range is displaced to the lower pressures.

Table 4 presents the values of the normalized surface potential V / V_0 for samples stored at pressure of 0.1mbar between two short circuited plate electrodes at different thickness values d for the air gap.

U_g, V	V / V_0				
	$d_1 = 0.10$mm	$d_2 = 0.28$mm	$d_3 = 0.84$mm	$d_4 = 1.69$mm	$d_5 = 3.00$mm
-350	0.47	0.44	0.46	0.45	0.47
-500	0.13	0.19	0.25	0.27	0.27
-650	0.05	0.07	0.11	0.14	0.19
-800	0.05	0.06	0.09	0.11	0.14
-950	0.04	0.06	0.07	0.09	0.11

Table 4. Normalized surface potential values V / V_0 (accuracy is better than 5%) for PP samples stored between two short-circuited plate electordes at the pressure of 0.1 mbar ($\log p / p_0 = -4$).

From the results presented in Table 4 the following conclusion can be made:
The final values of the surface potential at 0.1mbar decrease with the increase of the initial surface potential and grow with the increase of the air gap thickness.

3.1.5 Electrets obtained in a corona discharge and stored at low pressures in the absence of a second electrode

Two series of experiments with polypropylene samples were carried out. In the first one, the samples were charged in positive or negative coronas at room temperature, while in the second one they were charged in a positive or a negative corona at temperatures higher than the room one: 75°C and 90°C. The cleaning of the samples, their charging and the measurement of their surface potential were carried out as were described in 3.1.2. The samples were stored in a vacuum chamber in the absence of a second electrode, i.e. in an air gap tending to infinity.

Normalized surface potential V / V_0 vs normalized p / p_0 pressure for samples charged in a positive or a negative corona respectively at room temperature are presented in Fig. 6 and Fig. 7, while the samples in Fig. 8 and Fig. 9 are charged in a positive or a negative corona respectively at temperature higher than the room one. As the electrets charged at 75°C and 90°C have the same behaviour the only presented results are the ones obtained at 90°C.

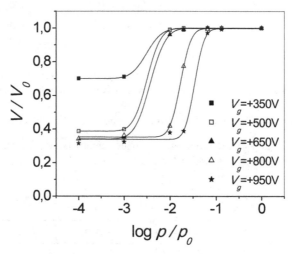

Fig. 6. Dependence of the normalized surface potential on normalized pressure for positively charged PP samples at room temperature. Infinite air gap.

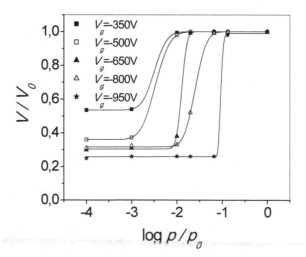

Fig. 7. Dependence of the normalized surface potential on the normalized pressure for negatively charged PP samples at room temperature. Infinite air gap.

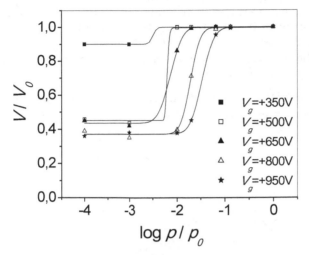

Fig. 8. Dependence of the normalized surface potential on the normalized pressure for positively charged PP samples at 90°C. Infinite air gap.

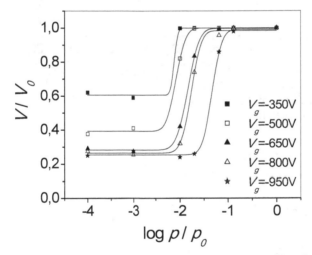

Fig. 9. Dependence of the normalized surface potential on the normalized pressure for negatively charged PP samples at 90°C. Infinite air gap.

Each point corresponds to the average value obtained by the measurement of 6 samples. The maximum deviation from the average values determined at 95% confidence level is 5%.

3.1.6 Time dependence of the surface potential of electrets obtained by corona charging and stored at low pressures in the absence of a second electrode

The storage time dependence on the decrease of the surface potential of PP electrets charged in negative corona at room temperature and at 90°C was studied. Voltage of U_c = -5kV was applied to the corona electrode and to the grid - U_g = -800V. A part of the electrets obtained was stored at pressure of 0.1mbar for 30 minutes and after they were taken out of the vacuum chamber, their surface potential had been measured for 30 days. The surface potential of the other electrets had been measured in time just after their charging without preliminarily storage at low pressure.

The time dependence of the normalized surface potential for PP samples negatively charged at room temperature is presented in Fig. 10 and in Fig. 11 – for PP samples charged at 90∘C.

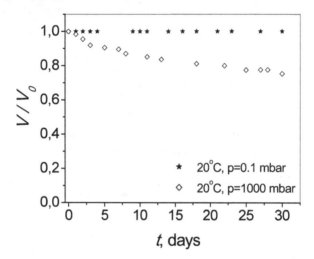

Fig. 10. Time dependence of the normalized surface potential for PP samples negatively charged at room temperature.

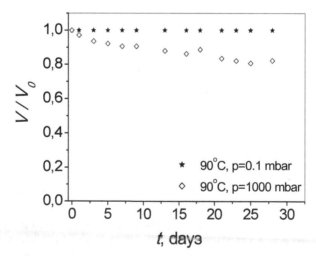

Fig. 11. Time dependence of the normalized surface potential for PP samples negatively charged at 90∘C.

For each figure point the measured standard deviation does not exceed 10% of the average value at 95% confidence level.

It was found that:

- For samples preliminarily stored at pressure of 0.1mbar for 30 minutes the surface potential, independently of the charging temperature, does not change over time ;
- For samples which have not preliminarily been charged at low pressure the surface potential value decreases to a certain point over time.

Hence, the preliminarily placement of the electrets at low pressure for a certain period of time can be used as an effective method for stabilizing the electrets charge depending on their applications.

3.1.7 Temperature dependence of the surface potential of corona charged electrets

The dependences of normalized surface potential on the temperature for negatively charged PP films were investigated. Voltage of U_c = -5kV was applied to the corona electrode and to the grid - U_g = -500V. A part of the electrets was charged at room temperature and the other electrets – at temperature 90ºC. The electrets obtained were preliminary stored for 30 minutes at two different pressures (0.1mbar and 1000mbar). After that the dependences of normalized surface potential on the temperature was measured.

The dependences of normalized surface potential on the temperature for negatively charged PP films are presented in Fig. 12.

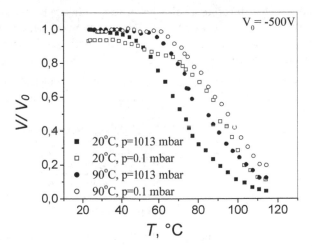

Fig. 12. Temperature surface potential decay curves after negative corona charge.

The results show that the curves of the samples which were preliminary stored at low pressure are shifted slightly to higher temperature as the shallow traps are liberated and the curves (Fig. 12) are shifted to higher temperature. When the temperature increases, the transport processes through the bulk will play a determinant role because of the increase of the injection of the carriers from the surface into the bulk and generating carriers in the bulk and the increase of their mobility.

3.2 Discussion

It is typical feature for the electret transducers that in most cases the electret is one-side metalized and is placed between two plate electrodes short-circuited by a resistor. The charged electret surface gives rise to an electric field in the air gap which is formed between the non-metalized electret surface and its adjacent electrode. This field can be found by an equation analogous to the one in (Palaia et al.):

$$E = \frac{\varepsilon V}{\varepsilon d + \varepsilon_1 L},$$ (1)

where E is the air gap field, ε is the electret relative dielectric permeability, $\varepsilon_1 = 1$ and d are the relative dielectric permeability and thickness of the air gap respectively, L - the electret thickness, V is the electret surface potential.

It is assumed in most of the papers that the electrets charge decay is due to gas discharges in the electret-electrode gap which depend on the gas pressure and the thickness of the gap itself according to Paschen's law (Knoll et al., Schaffert R.).

For dry air the Paschen's curve has a minimum at pd =6.65mbar.mm corresponding to a breakdown voltage of 360V (Schaffert R.). If the air gap field creates a voltage less than the breakdown voltage for the particular gas, gas discharges will not occur in the gap. When the air gap field creates a voltage equal or higher than the breakdown voltage, a discharge will occur and the electret charge will consequently decrease. The discharge will continue until the voltage across the air gap is reduced to a value below the minimum breakdown voltage for the respective value of the product pd and will not depend on both the size and the polarity of the initial electret surface potential as well as on the air gap thickness. Therefore, the minimal voltage value in the gap at which a discharge can be begun is 360V (Schaffert R.).

The values of the fields in and the voltages across the electret-electrode gap calculated for PP samples stored between two short-circuited electrodes with air gap thickness of 0.10mm are presented in Table 5.

Surface potential V, V	400	500	650	800	950
Electric field in the air gap E, kV/cm	36.7	45.9	59.6	73.4	87.1
Voltage in the air gap U, V	367	459	596	734	871

Table 5. Electric fields and voltages in a 0.10mm air gap at different surface potential values.

Analogous calculations have been made for the other gaps used in our experiments. In all cases the voltages in the gaps do not exceed the grid surface potential. At atmospheric pressure and various thicknesses of the air gap used in our experiments the pd values change from pd_{min} (d = 0.1mm) = 100 mbar.mm to d_{max} (d = 3mm) = 300 mbar.mm, which corresponds to breakdown voltages higher than 950V. Hence, the calculations show that at atmospheric pressure at different air gap thicknesses d the voltage across the air gap is less than the breakdown voltage according to the Paschen's law (Schaffert R.) for all thicknesses d studied and for all grid voltages U_g. Therefore, there will be no discharge.

When the pressure in the vacuum chamber decreases, the breakdown voltage also decreases, while the voltage across the air gap created by the electret practically does not change. Hence, the voltage across the air gap might be found higher than the one following from the Paschen's law and it may result in a discharge.

At pressure of 0.1mbar and air gap thicknesses from 0.10mm to 3.00mm, the pd product values lay in the (0.01-0.30)mbar.mm interval. These values are much lower than the 6.65mbar.mm value where a minimum of the Paschen's curve is observed and they correspond to a higher breakdown voltage than the minimal one of 360 V. Therefore, the air gap pressure curve will cross the Paschen's curve and the surface potential is expected to decrease to the minimal value of 360V corresponding to pd =6.65mbar.mm. However, the results obtained show that the surface potential has decreased to values considerably lower than the minimal pressure value following the Paschen's law. This can be seen from the curves presented in Fig.13.

The dependence of the minimal value V_{min} to which the surface potential at pressure 0.1mbar has decreased as a function of the initial surface potential V_0 for the different air gap thicknesses is presented in Fig. 13.

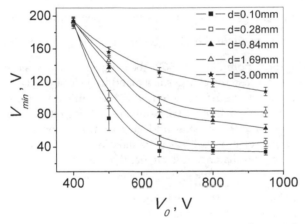

Fig. 13. Dependence of the minimum surface potential V_{min} on the initial surface potential V_0 at pressure of 0.1mbar for different air gaps.

Each figure point corresponds to an average value obtained by the measurement of 6 samples. The maximum deviation from the average values determined at 95% confidence level is 5%.

It can be seen from Fig. 13 that the surface potential decreases to a value lower than 360V for all values of the initial surface potential and all air gap thicknesses. With the air gap decrease and the initial surface potential increase the minimal value to which the electret surface potential decays is reduced. It decreases to a value of 33.5V at a gap of 0.10mm and initial surface potential 950V. This value of the surface potential corresponds to an electret-electrode gap voltage of 30.7V. This is considerably lower than the minimal value according to the Paschen's law. On the other hand, it is supposed that as the initial surface potential decreases the minimum surface potential V_{min} tends to a value approximately equal to an initial surface potential of 300V. Some additional experiments were carried out in order to verify these values. Several series of samples were charged to an initial surface potential of 280V, 300V, 320V and 350V respectively. The electrets were stored at a pressure of 0.1mbar for 30 minutes. The gap between the electret surface and the upper electrode was 0.28mm. It was established that for samples charged to initial values of the surface potential 280V, 300V and 320V the normalized surface potential was 0.98 and for the ones charged to 350V – 0.61. Hence, if the initial surface potential value is approximately 300V, no surface potential decay takes place.

The analysis of the results obtained in different experiments shows that:

- First, discharges depend on the corona polarity and the charging temperature;
- Second, the surface potential decay process is analogous for both samples stored between two short-circuited electrodes with different air gap thicknesses and the ones with the absence of a second electrode;
- Third, the surface potential decreases to a value lower than the minimal breakdown voltage according to the Paschen's law.

If we assume that the charges obtained due to the corona discharge are located on the sample surface, we can calculate the charge surface density by the equation $\sigma = \dfrac{\varepsilon_0 \varepsilon}{L} U_k$.

When the electret surface potential is 950V then the surface charge density is equal to $\sigma = 9.25 \times 10^{-8} C/cm^2$. In this case the charges are found to be located at a distance thereabouts 13nm apart. When the electret surface potential is 350V this distance is thereabout 21nm. Therefore, there is a discrete charge distribution on the electret surface and it results in a nonuniform electric field having normal and tangential components, close to the surface (Neugebauer H., Pisanova et al.). At a distance three or four time larger than the distance between the charges, i.e. at a distance not greater than 80nm from the charged surface, the field is in fact homogeneous (Pisanova et al., Feynman et al.).

As it is shown in (Schaffert R.) the distribution of the electric field close to the charged surface is practically the same in both the presence and the absence of a second electrode. These results give an explanation to the similarity found in those cases when the electrets were stored in the absence of a second electrode (Mekishev et al., 2007, Viraneva et al.). It is established (Giacometti et al.) that during the corona discharge in the air, at atmospheric pressure, different types of ions are deposited on a sample while charging in a corona discharge depending on the corona polarity. In the case of a positive corona the ions are mainly $H^+(H_2O)_n$ and the ones for a negative corona - CO_3^- (Giacometti et al.). Those ions are bound in traps of various depths and are released from them depending on the surrounding conditions. It was found in (Yovcheva et al.) by the X-ray photoelectron spectroscopy method that oxygen content in negatively charged samples is higher than that in positively charged samples. The oxygen content for non-charged samples is intermediate if compared to the values for the negatively and positively charged ones. It is assumed that the different oxygen content in the various cases is in consequence of various sorption processes on the sample surface. These results make us suppose that the main process responsible for the surface potential decay can be associated with desorption of charged species from the electret surface under the influence of its own electric field. These might be ions deposited on the surface or groups in which the ions have given their charge away.

In the published papers one can see different models for explaining the adsorption (desorption) mechanisms. The experimental results were fitted by the Friondlih and Lengmuir equations in order to estimate how well these models describe the behavior of the surface potential of electrets stored at low pressure. The dependence of the normalized surface potential on the pressure for PP samples charged in negative corona at room temperature and stored between two plate electrodes with air gap thickness 3.00mm is presented in Fig. 14. One can see from Fig. 14 that the equations used do not describe the results obtained well enough. Analogous graphs were plotted for the other gaps (smaller than 3mm). It was established that for smaller gaps the fitting accuracy sharply decreases.

The dependences of the normalized surface potential on pressure for PP samples charged in a negative corona at room temperature and temperature 90°C to the initial surface potential value of 650V and stored in the absence of a second electrode are presented in Figs. 15, 16.

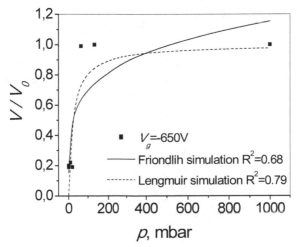

Fig. 14. Dependence of the normalized surface potential on pressure for negatively charged PP samples at temperature of 20°C, fitted by the Friondlih and Lengmuir equations. The air gap thickness is 3.00mm.

Analogous graphs were plotted for the samples charged in a positive corona.

The results presented in Fig.14-16 show that irrespective of the corona polarity and the charging temperature the Friondlih and Lengmuir equations do not describe very well the results obtained in our experiments as the determination factor is $0.36 \leq R^2 \leq 0.85$. That is why we assumed that the desorption process might run together with another process like charge movement along the surface, for example. Our assumption is based on the results reported in various papers (Protodyakonov et al., Baum et al., Karmazova et al., Atkinson et al., 1976, Atkinson et al., 1980, I, Atkinson et al., 1980, II) and some additional experiments we have carried out with electrets with an island charge distribution. It is reported in

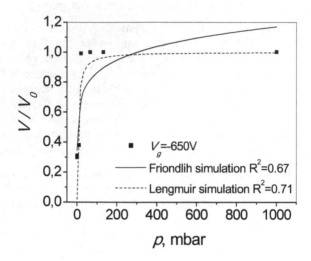

Fig. 15. Dependence of the normalized surface potential on pressure for PP samples charged in negative corona, at 20°C fitted by the Friondlih and Lengmuir equations. Infinite air gap.

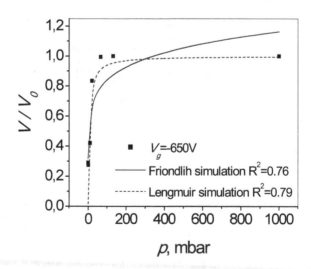

Fig. 16. Dependence of the normalized surface potential on pressure for PP samples charged in negative corona at 90°C, fitted by the Friondlih and Lengmuir equations. Infinite air gap.

(Protodyakonov et al.) that under certain conditions the adsorbate molecules located on the surface can be transferred along the surface without leaving it. The paper (Karmazova et al.) studies the influence of storage time on the surface potential decay for PP and PTFE samples with uniform and island charge distribution charged in corona discharge at room temperature and atmospheric pressure. It is shown that for samples with island charge distribution the surface potential decay is much smaller than for the ones with uniform charge distribution.

We carried out an experiment analogous to the one described in (Karmazova et al.): to charge PP samples in positive or negative corona we used a set-up for obtaining an island charge distribution described in (Karmazova et al.).

This set-up contains a grounded plate electrode, a metal mask (instead of a grid) with a numerous apertures of a definite size and distribution of the apertures and a corona electrode (a needle). The metal mask functions as a control electrode. During the charging process the polymer film was placed on the grounded plate electrode and the mask with the apertures was put on the film. The voltage U_c =±5kV was applied to the corona electrode and U_g =±650V – to the metal mask. The shape, the size and the distribution of the apertures in the mask define the shape, the size and the distribution of the charge on the electret surface (Fig.17).

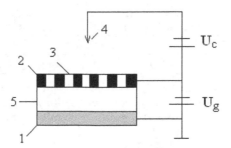

Fig. 17. Set-up for obtaining electrets with island charge distribution: 1 – grounded plate electrode; 2 – metal mask; 3 – apertures; 4 – corona electrode; 5 – electrets; U_c - corona voltage; U_g - metal mask voltage.

After obtaining the electrets the metal mask was removed from the charged film. The electret surface potential was measured by the method of vibrating electrode with compensation. Then the electrets were placed in a vacuum chamber for 30 minutes at various low pressures. After the samples were taken out of the vacuum chamber, their surface potential was measured again and the normalized surface potential V / V_0 was calculated. The dependence of the normalized surface potential V / V_0 on the normalized pressure p / p_0 for PP samples with island charge distribution is presented in Fig. 18.

Each graph point corresponds to the average value obtained by the measurement of 6 samples. The maximum deviation from the average values determined at 95% confidence level is 5%. The results for samples with island charge distribution stored at pressure of 0.1mbar are compared to the ones for samples with uniform charge distribution. These results are presented in Table 6.

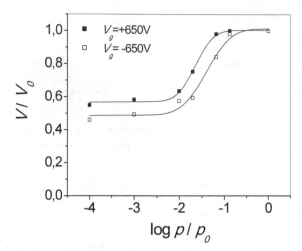

Fig. 18. Dependence of the normalized surface potential on normalized pressure for PP samples with island surface charge distribution.

Grid voltage	V / V_0	
	Uniform charge distribution	Island charge distribution
+650V	0.34	0.55
-650V	0.30	0.46

Table 6. Normalized surface potential V/V_0 values (with an error not exceeding 5%) for samples charged in positive or negative corona with uniform or island surface charge distribution. The samples were stored at pressure of 0.1mbar with the absence of a second electrode.

One can see from the table that when the samples were stored at pressure of 0.1mbar for 30 minutes the ones with an island charge distribution have higher surface potential values than the samples with uniform charge distribution, independent of the corona polarity. These results show that the assumption for the movement of the charges on surface seems completely plausible. The absorption and resorption currents in FEP teflon samples with 4.475 cm diameter are investigated in (Atkinson et al., 1976). One of the sample surfaces is completely metalized and an aluminum electrode with diameters ranging from 0.32cm to 2.54cm is evaporated in the central part of the second surface. The surface completely metalized is usually grounded and a definite voltage is applied to the upper electrode. It is established that there is a charge movement on the polymer surface between the metal electrode end and the non-metalized polymer surface. This movement is the main component of the absorption and resorption currents. The surface components of the absorption and resorption currents in vacuum for different polymers were later studied in (Atkinson et al., 1980 I). It was established for FEP teflon, polystyrene, polytetrafluorethylene and polythene that the charge movement along the polymer surface is the main component of these currents. Baum and coauthors were studied the surface charge decay in PET films (Baum et al.). Using a corona discharge (positive or negative corona) a charged spot with a diameter of 0.5cm is created on one of the surfaces. It is established that the charges deposited by the corona source move laterally according to a diffusion law. The time dependence of the absorption currents is deduced in (Atkinson et al., 1980 II) on the assumption that these currents are entirely due to the transverse movement of charges caused by a concentration gradient of charge carriers. An equation combining linear desorption with surface diffusion is proposed for the description of some desorption processes in (Keltzev N.).

Our experimental results are well described by an equation analogous to the one in (Keltzev N.).

$$\theta = a + \frac{1}{2}b\left(1 + erf\left(\frac{x-c}{\sqrt{2}d}\right)\right),$$ (2)

where $\theta = V/V_0$ is the normalized surface potential, $x = p/p_0$ is the normalized pressure and a, b, c and d are constants depending on the charging conditions. Values of the parameters a, b, c and d for PP samples charged in positive or negative corona to an initial surface potential value of 800V are presented in Table 7.

Corona polarity	a	b	c	d
positive	0.3536± 0.0065	0.6439± 0.0085	-1.7702± 0.0076	0.1809± 0.0109
negative	0.3167± 0.0034	0.6835± 0.0053	-1.5964± 0.0088	0.1932± 0.0143

Table 7. Values of the parameters a, b, c and d obtained by fitting the equation 2.

The equation (2) is analyzed by using the $erf\left(\frac{x-c}{\sqrt{2}d}\right)$ values for different cases (Abramovich et al.). It should be noted that $d > 0$.

Case 1: $\left|\frac{x-c}{\sqrt{2}d}\right| \gg 1$

a. $x - c > 0$

In this case $erf\left(\frac{x-c}{\sqrt{2}d}\right) = 1$ and from equation (2) it follows that: $\theta = a + b = $ const

Therefore, at high values of the ratio p/p_0, $\theta = a + b = 1$ and the surface potential practically remains equal to the initial surface potential, i.e. $V \approx V_0$. In this interval of the ratio p/p_0 no changes of the surface potential are observed.

b. $x - c < 0$

In this case $erf\left(\frac{x-c}{\sqrt{2}d}\right) = -1$ and from equation (2) it follows that:

$$\theta = a = const$$

Therefore, if $x < c$, the surface potential remains constant. Then the parameter a has the meaning of the minimum value of the normalized surface potential and b is the difference between the maximum and minimum values of the normalized surface potential.

Case 2: $\left|\dfrac{x-c}{\sqrt{2d}}\right| \ll 1$

In this case $erf\left(\dfrac{x-c}{\sqrt{2d}}\right)$ develops into a series and only the first linear term from the expansion can be considered:

$$erf\left(\frac{x-c}{\sqrt{2d}}\right) = \frac{2}{\sqrt{\pi}}\frac{x-c}{\sqrt{2d}}$$

and from equation (2) a linear dependence of θ on x is obtained:

$$\theta = A + Bx$$

where $A = a + \dfrac{b}{2} - \dfrac{bc}{\sqrt{2\pi d}}$ and $B = \dfrac{b}{\sqrt{2\pi d}}$.

If $x = c$, $erf\left(\dfrac{x-c}{\sqrt{2d}}\right) = 0$ and from equation (2) it follows that: $\theta = a + \dfrac{b}{2}$

Therefore, c is the midpoint of the range p / p_0 , within which the surface potential decay occurs.

When $x = c \pm \sqrt{2d}$, $erf\left(\dfrac{x-c}{\sqrt{2d}}\right) = \pm 0.843$, i.e. the decay range for the normalized surface potential is $\left(c - \sqrt{2d}, c + \sqrt{2d}\right)$.

All curves presented in Fig. 4-9 are described very well by equation (2) with a coefficient of determination $0.98 \le R^2 \le 1.00$. It was established that charging the electrets to various surface potential values leads to a curve displacement. The higher the initial surface potential the higher the pressure at which the sharp decay occurs. Therefore, there are two factors that influence the surface potential decay – the pressure at which the electrets have been stored and the initial surface potential to which they have been charged. The ratio p / V_0 is the main factor determining the surface potential decay range. Therefore, each group of experiments can be described by a generalized curve.

The dependences of the normalized surface potential V / V_0 on the normalized ratio $\left(p/V_0\right)^* = \left(p/V_0\right)\left(p_0/V_0^*\right)^{-1}$ for positively and negatively charged PP samples are presented in Fig. 19 and Fig. 20.(Mekishev et al., 2005)

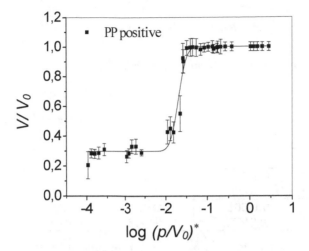

Fig. 19. Dependence of the normalized surface potential V / V_0 on the normalized ratio $(p/V_o)^*$ for positively charged PP samples.

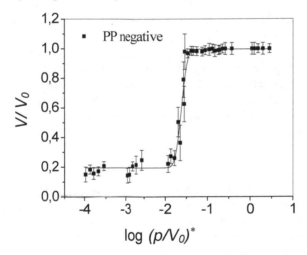

Fig. 20. Dependence of the normalized surface potential V / V_0 on the normalized ratio $(p/V_o)^*$ for negatively charged PP samples.

The symbol p_0 stands for the atmospheric pressure and $V_0^* = 1000$ V. For each graph point (Fig. 19, 20) the calculated standard deviation does not exceed 10% of the average value at 95% confidence level. It must be noted that the errors are smaller within the higher values of the normalized ratio $(p/V_o)^*$ and are higher in the range of the sharp decay.

It was found out that the results obtained can be described by an equation analogous to equation (2). The results presented so far are described very well with a coefficient of determination $0.98 \leq R^2 \leq 1.00$ by the following equation:

$$\theta = a^* + \frac{1}{2}b^*\left(1 + erf\left(\frac{x^* - c^*}{\sqrt{2}d^*}\right)\right),$$ (2a)

where $\theta = V/V_0$ is the normalized surface potential,

$x^* = \log(p/V_0)^* = \log\left[(p/V_0) \cdot (p_0/V_0^*)^{-1}\right]$ and a^*, b^*, c^* and d^* are constants depending of

the charging conditions. Values of this parameters for positively and negatively charged PP samples are presented in Table 8.

Corona polarity	a^*	b^*	c^*	d^*
positive	0.2977± 0.0153	0.7037± 0.0196	-1.6949± 0.0176	0.1508± 0.0214
negative	0.1950± 0.0167	0.8017± 0.0218	-1.6145± 0.0111	0.1051± 0.0159

Table 8. Values of the parameters a^*, b^*, c^* and d^* obtained by fitting the PP results by equation 2a.

It is seen from the results presented in Fig. 19, 20 and Table 8 that the minimum value of the surface potential after the sharp decrease is higher for the positively charged samples (higher values of parameter a^*). The values of the ratio p/V_0 for the midpoint of the region of the sharp surface potential decay were calculated using the values of parameter c^* from Table 8. The following values were obtained: 0.019mbar/V and 0.024mbar/V for the positively and negatively charged PP samples. Therefore, the ratio p/V_0 for the midpoint of the region, within which the sharp surface potential decay occurs, allows calculating the pressure at which the sharp electret surface potential decay will be observed if their initial surface potential is known.

4. Conclusion

The influence of low pressure on the behaviour of corona charged PP electrets was studied. The PP films were charged in a positive or a negative corona at different charging temperatures of T=20°C and T=90°C and were stored in a vacuum chamber under various low pressures. It was found that the charging time of 30 minutes does not influence the surface potential of the PP films. It was established that the surface potential decreases within the first minute and that further storage at the reduced pressure does not bring any change.

The analysis of the results obtained in different experiments shows that:

First, the surface potential decay depends on the corona polarity and the charging temperature; Second, the surface potential decay process is analogous for both samples stored between two short-circuited electrodes with different air gap thicknesses and the ones with the absence of a second electrode; Third, the surface potential decreases to a value lower than the minimal breakdown voltage according to the Paschen's law.

It is supposed that the main process responsible for the surface potential decay can be associated with desorption of charged species from the electret surface under the influence of its own electric field accompanied by surface diffusion. An equation combining linear desorption with surface diffusion is proposed for the description of the surface potential decay. This equation described very well the results obtained with a coefficient of determination $0.98 \le R^2 \le 1.00$. It is proposed a generalized curve describing the dependence of normalized surface potential on the ratio p / V_0. From this curve it is possible to calculate pressure at which sharp decay will occur if the dependence of the surface potential on the pressure and the initial surface potential are known.

The preliminarily placement of the electrets at low pressure thereabout 1mbar for a certain period of time can be used as an effective method for stabilizing the electrets charge depending on their applications.

5. Acknowledgment

The authors gratefully acknowledge the financial support of National Science Fund, Bulgaria. The authors also gratefully acknowledge the financial support from the Biosupport project No 245588 (7th FWP).

6. References

Abramovich, M. & Stegun, I. (1964). (eds.) *Handbook of Mathematical Functions*, Applied Mathematics, Series 55 (National bureau of standarts)

Adamson, A. & Gast, A. (1977). *Physical chemistry of surfaces*, John Wiley @ Sons, Inc., New York-Chichester-Weinheim-Brisbane-Singapoore-Toronto

Atkinson, P. & Fleming, R. (1980). Surface component of vacuum absorption and resorption currents in polymers: I. Origin and magnitude. *Journal Phys. D: Appl. Phys.*, Vol.13, pp. 625-638

Atkinson, P. & Fleming, R. (1980). Surface component of vacuum absorption and resorption currents in polymers: II Surface charge accumulation. *Journal Phys. D: Appl. Phys.*, Vol.13, p. 639-653

Atkinson, P. & Fleming, R. (1976). Origin of absorption and resorption currents in the co-polymer poly(hexafluoropropylene-tetrafluoroethylene). *Journal Phys. D: Appl. Phys.*, Vol.9, pp. 2027-2040

Baum, E.; Lewis, T. & Toomer, R. (1978). The lateral motion of charge on thin films of polyethylene terephtalate. *Journal Phys. D: Appl. Phys.*, Vol.11, pp. 963-977

Draughn, R. & Catlin A. (1968). Effect of Low Pressure on Surface Charge of Electrets. *Journal Electrochemical Society: Solid State Science*, Vol.115, №4, pp. 391-394

Feynman, R.; Leighton, R. & Sands, M. (1964). *The Feynman lectures on physics*, Addison-Wesley Publishing Company, Inc., Reading, Massachusetts, Palo Alto, London

Gang-Jin, C.; Hui-ming, X. & Chun-feng, Z. (2004). Charge dinamic characteristics in corona-charged polytetrafluoroethylene film electrets. *Journal of Zhejiang University Science*, Vol.5, №8, pp. 923-927

Giacometti, J.; Fedosov, S. & Costa, M. (1999). Corona Charging of Polymers: Recent Advances on Constant Current Charging. *Brazilian Journal of Physics*, Vol.29, №2, pp. 269-279

Goel, M. (2003). Electret sensors, filters and MEMS devices: New challenges in materials research. *Current science*, Vol.85, №4, pp. 443-453

Gross, B. (1969). Discussion of Effect of Low Pressure on Surface Charge of Electrets. *Journal of the Electrochemical Society*, Vol.116, №6, pp. 874-874

Gubkin, A. & Skanavi, G. (1961). *Sov. Phys. – Solid state*, Vol.3, pp. 215

Jeda M., Sawa G. & Shinohara U. (1967). A decay process of surface electric charges across polyethylene film. *Japanese Journal of Applied Physics*, Vol6, №6, pp. 793-794

Karmazova, P. & Mekishev, G. (1992). Electrets with Island Surface Charge Distribution. *Europhysics Letters*, Vol.19, №6, pp. 481-484

Keltzev, N. (1984), *Bases of Adsorption Technique*, Chimie, Moskow, (in Russian)

Knoll, M.; Ollendorff, F. & Rompe, R. (1939). *Gasentladungstabelen*, Springer-Verlag, Berlin, pp. 84

Mekishev, G.; Yovcheva, T. & Viraneva, A. (2007). Investigation of PP and PTFE film electrets stored at low pressure. *Journal of Non-Crystalline Solids*, Vol.353, pp. 4453–4456

Mekishev, G.; Yovcheva, T.; Gencheva, E. & Nedev, S. (2005). Study of electrets stored at pressure lower than atmospheric. *Journal of Electrostatics*, Vol.63, pp. 1009-1015

Mekishev, G.; Yovcheva, T.; Guencheva, E. & Nedev, S. (2003). On the charge decay in PPelectrets stpred at pressures lower than atmospheric. *J.Mater.Sci.: Materials in electronics* Vol.14, pp. 779-180

Mohmeyer, N.; Behrendt, N.; Zhang, X.; Smith, P.; Altstadt, V.; Sessler, G. & Schmidt, H. (2007). Additives to improve the electret properties of isotactic polypropylene. *Polymer*, Vol.48, pp. 1612-1619

Nalwa, H. (1995). *Ferroelectric polymers*, Marcel Dekker, Inc., New York

Neugebauer, H. (1964). Electrostatic Fields in Xerography. *Applied Optics*, Vol.3, №3, pp. 385-393

Ono, R.; Nakazawa, M. & Oda T. (2004). Charge storage in Corona-Charged Polypropylene Films Analysed by LIPP and TSC Methods. *IEEE Transaction on Industry Applications*, Vol.40, №6, pp. 1482-1487

Palaia, F. & Catlin A. (1970). Electret Behavior at Low Pressure. *The Journal of chemical physics*, Vol.52, №7, pp. 3651-3654

Pisanova, E. & Mekishev, G., unpublished data

Protodyakonov, I. & Siparov S. (1985). *Mechanics of adsorption processes in system gas-solids*, Nauka, Leningrad, pp. 299

Reedyk, C. & Perlman, M. (1968). Method for measurement of surface charge on electrets. *Journal of the Electrochemical Society*, Vol 115, №1, pp. 49-51

Ribeiro, P.; Giacometti, J.; Raposo, M. & Marat Mendes, J. (1991). Effect of the air humidity on the corona polarization of β - PVDF films. *7th Int. Symp. on Electrets*, pp. 322-327

Schaffert, R. (1965). *Electrophotography*, The Focal Press, London, New York

Sessler, G. & Gerhard-Multhaupt, R. (1999). *Electrets*, third edition, Laplacian Press, Morgan Hill, California`

Sessler, G. (1980). *Electrets*, Springer - Verlag, Berlin – Heidelberg, New York

Shepard, G. & Stranathan J. (1941). Effect of Pressure on the Surface Charge of an Electret. *Physical Review*, Vol.60, pp. 360-361

Viraneva, A.; Yovcheva, T.; Gencheva, E. & Mekishev, G. (2008). Study of PET corona electrets at atmospheric and lower pressures. *Journal of Optoelectronics and Advanced Materials*, Vol.10, №2, pp. 302-305

Wild, J. & Stranathan, J. (1957). Influence of External Treatments on Electret Behavior. *The Journal of Chemical Physics*, Vol.27, №5, pp. 1055-1059

Yovcheva, T.; Avramova, I.; Mekishev, G. & Marinova, T. (2007). Corona-charged polypropylene electrets analyzed by XPS. *Journal of Electrostatics*, Vol.65, pp. 667-671

Zhongfu, X.; Yuda, W.; Guamao, Y. & Ximin S. (1991). Corona Charging at Elevated Temperature and Charge Transport for Mylar PETP Foils. *7th Int. Symp. on Electrets*

11

Acoustic and Dielectric Properties of Polypropylene-Lignocellulosic Materials Composites

Ewa Markiewicz[1], Dominik Paukszta[2] and Sławomir Borysiak[2]
[1]Institute of Molecular Physics, Polish Academy of Sciences, Poznań,
[2]Institute of Technology and Chemical Engineering,
Poznan University of Technology, Poznań,
Poland

1. Introduction

In recent years, one can observe the tendency to replace the thermoplastic polymers by the composite materials in several branches of industry, e.g. automotive and building engineering (Peijs, 2003), aviation and packaging industry (Bledzki & Gassan, 1999). The composite materials obtained by reinforcement of the polypropylene with lignocellulosic fillers are known to show improved mechanical and physical properties in comparison with the pure propylene (Averous & Le Digabel, 2006; Bhattacharyya et al., 2003; Mohanty et al., 2000). Composites made from polypropylene and wood fibre are characterized by significantly higher stiffness than unreinforced polypropylene (Bhattacharyya et al., 2003). The loading of the polypropylene with rice husk powder increases Young's modulus and flexural modulus of the composite, compared with those of the polypropylene (Hattotuwa et al., 2002). The studies of the fire behavior (Borysiak et al., 2006) revealed a significant decrease of such an essential parameter as the heat release rate (HRR) peak, especially low value of HRRmax in comparison with those of the polypropylene. Moreover, very important feature of the composites with lignocellulosic materials is their partial biodegradability as the filler materials come from natural resource. In this chapter we would like to point out to improved acoustic and dielectric properties of the polypropylene-lignocellulosic materials composites in comparison with the pure polypropylene based on our measurement results.

Nowadays, technical progress in manufacturing the modern equipment, generating higher sound pressure, implies the need to search for new sound absorbing materials to improve the human comfort today. The ascending requirements related to the construction materials absorbing the undesired noise occur mainly in the automotive industry and building. The commonly used and unporous materials as ceramic tile, concrete, cement, fiberboard and playwood are characterized by weak sound absorption properties with the sound absorption coefficient bellow 5% in the frequency range from 125 Hz to 8000 Hz (Tiwari et al., 2004; Yang et al., 2003). The sound absorption capacity of the environment is usually corrected by the sound absorbance systems made from glass wool, foam (metals and polyurethanes), rubber, mineral fibres and their composites. Although the sound absorption can be significantly increased due to installation of these traditionally applied materials, they cause environmental pollution and pose danger to human health. Recent tendency towards the environmental protection stimulates the utilization of natural materials as sound absorbers, e. g. random cut rice straw (Yang et al., 2003), coconut coir (Nor et al., 2004), bamboo (Liu & Hu, 2008) and tea-leaf-fibre (Ersoy & Kucuk, 2009). We propose the use of composites made from polypropylene and lignocellulosic material derived from hemp, flax, beech wood and rapeseed straw as promising sound absorbers (Markiewicz et al., 2009).

Combination of the polymer and the lignocellulosic material results in new dielectric properties of the composite. The proper formation of the dielectric properties of the composites is very important in the field of their application, particularly when they are designed as electronic packaging. In this case, the electrical parameters of the microelectronic devices, such as signal attenuation, propagation velocity, and cross talk, are influenced by the dielectric permittivity value ε', dielectric losses ε'' and their temperature stability in a wide frequency range (Pecht et al., 1999; Chung, 1995; Subodh et al., 2007). On one hand the permittivity ε' should not be low because of the demand for the miniaturization of the device but on the other hand it cannot be too large in order to enable the high signal propagation speed. The signal delay T_d propagated through the metal embedded in the packaging material is determined by the dielectric permittivity ε' according to the formula (Tummala, 1991):

$$T_d = \sqrt{\frac{\varepsilon'}{c}} \qquad (1)$$

where c is the elastic coefficient. It is evident that high dielectric permittivity ε' reduces the signal propagation speed. Similarly, in the case of application of the composite as the substrate in sensor of acoustic surface waves the signal propagation speed v is reduced due to high dielectric permittivity ε' in accordance with the relationship:

$$v = \sqrt{\frac{c + \dfrac{e^2}{\varepsilon'}}{\rho}} \cdot \sqrt{1 - \beta^2} , \qquad (2)$$

where e denotes the piezoelectric coefficient, ρ stands for the density and β is the factor of decreasing the signal amplitude inside the substrate material (Soluch, 1980). In this chapter, the relationship between the dielectric permittivity ε' of the composite and the volume fraction of the lignocellulosic material is established. The effect of temperature variation from 150 to 450 K on the dielectric spectrum of polypropylene and the composites was investigated in the frequency range 100 Hz to 1MHz.

2. Methods of sample preparation

Polypropylene-lignocellulosic materials were prepared from the following materials:

- isotactic polypropylene (PP) type Malen F-401 (melt flow rate $MFR_{230/2.16}$ = 2.4 -3.2g/10 min, isotacticity 95 %), produced by Basell Orlen Polyolefins (Poland) was used as a matrix for preparation of the composites;

- lignocellulosic materials derived from hemp, flax, beech, pine, rapeseed straw were used as filling materials.

Number	Kind of material	Density [kg/m³]
No. 1	Polypropylene (PP)	881.8
No. 2	PP+40% of crumble hemp plant	872.8
No. 3	PP+40% of long hemp fibres	927.9
No. 4	PP+40% of long flax fibres	934.6
No. 5	PP+40% of rapeseed straw Kaszub	918.8
No. 6	PP+40% of crude beech wood	803.5
No. 7	PP+20% of crude beech wood	921.0
No. 8	PP+20% of beech modified with succinic anhydride	920.7
No. 9	PP+20% of crude pine wood	850.9
No. 10	PP+20% of pine modified with succinic anhydride	974.1
No. 11	PP+30% of crude rapeseed straw Kaszub	889.6
No. 12	PP+30% of rapeseed straw Kaszub modified with acetic anhydride	904.0
No. 13	PP+30% of crude rapeseed straw Californium	924.8
No. 14	PP+30% of rapeseed straw Californium modified with acetic anhydride	969.9
No. 15	PP+25% of short hemp fibres	862.5
No. 16	PP+25% of hemp shivers	911.3
No. 17	PP+25% of short flax fibres	883.0
No. 18	PP+25% of flax shivers	943.1
No. 19	PP+30% of crude beech wood	922.1
No. 20	PP+30% of mercerized beech wood	922.0
No. 21	PP+30% of beech modified with maleic anhydride	921.9
No. 22	PP+30% of crude pine wood	861.1
No. 23	PP+30% of mercerized pine wood	895.4
No. 24	PP+30% of pine modified with maleic anhydride	901.3
No. 25	PP+30% of mercerized rapeseed straw Kaszub	902.5
No. 26	PP+30% of mercerized rapeseed straw Californium	950.7

Table 1. Specification of the samples investigated

Two different methods were used to make the composites. The first one consisted in mixing crumble lignocellulosic materials with polypropylene granulate in different proportion (20 – 40 wt. % of natural component). After that, the extrusion was carried out using a "Fairex"

(McNell Akron Repiquetn, France) single-screw extruder, L/D=25. The composite material was obtained in a granulate form (*Polish Patent* 186577, 2004). The composite granulates were melted in mould between heating plates at the temperature of 200°C under load of 3000 kG to obtain the samples required for the experiments.

The composites containing the long hemp and flax fibres were produced in a different way. A technique of hydraulic pressing at temperature 200°C under load of 3000 kG (*Polish Patent* 190405, 2005).

Finally, the samples took the shape of discs. Table 1 specifies all the samples prepared.

3. Acoustic properties of the polypropylene reinforced with lignocellulosic

3.1 Effect of reinforcement of polypropylene with lignocellulosic materials on the acoustic properties

The sound absorptive power of a given material sample is characterized by the sound absorption coefficient α defined as the ratio of the acoustic wave energy E_a absorbed by the sample to the total energy E_i incident on the sample:

$$\alpha = \frac{E_a}{E_i} \, . \tag{3}$$

Generally, the composites are known to exhibit better sound absorption than the homogenous materials. The fact results from the additivity of all kinds of acoustic energy losses (Epstein & Carhart, 1953; Vinogradov, 2004). The sound wave propagated through the inhomogeneous medium interacts with a great number of suspended particles, which differ by the density, compressibility and thermophysical parameters from the matrix. This leads to the additional acoustic energy losses compared to that in the matrix. The property of the additivity allows to express the sound absorption coefficient of the composite α as a sum of four components:

$$\alpha = \alpha_0 + \alpha_F + \alpha_H + \alpha_S \, , \tag{4}$$

where: α_0 - the coefficient of the matrix, α_F - the coefficient due to friction between filler particles and the matrix, α_H - the coefficient related to the heat exchange between the particles and the matrix and α_S - the coefficient caused by the decay of the acoustic wave in forward direction due to scattering by the particles. The results of the experimental studies by I. S. Kol'tsova et al. (Vinogradov, 2004) show that the sound absorption due to the scattering play an important role when the particle sizes are comparable or larger than the sound wave length. Thus, in the case of interaction between an acoustic waves of low frequency and the particles of micrometer/millimetre dimensions, the losses in acoustic energy due to friction and interfacial heat exchange play the main role. The different densities of the particles and the matrix are the reason for which the sound wave induced motions of both compounds can be considered as the separate ones with the friction existing between them. In real media, viscous forces arise balancing the motions of the particles and the matrix and giving rise to the sound absorption. When the heating coefficients on compression of both components are different, the effect of the variable sound pressure on the composite results in heat exchange between the components. At a macroscale, the process of compression and expansion proceeds adiabatically. However, at a microscale, i. e. in the scale of particle sizes, the process is nonadiabatic with the degree of heat exchange dependent on the frequency. At low frequencies, the temperature difference between the particles and the matrix has time to balance and the process is microscopically isothermal. In the higher frequency range, the process follows adiabatically at a microscale because balancing does not occur. The heat transfer through the filler particle - matrix interface is the reason for acoustic energy absorption (Vinogradov, 2004). Taking into account the above considerations, it can be stated that the increased sound absorption of the composites results from friction and interfacial heat exchange. This is an isothermal process, since the filler particles are of several millimetres in length and bellow millimetre in width.

The numerical calculation related to the elastic wave propagation in anisotropic media was initiated by Biot in 1955 (Biot, 1955, 1956a, 1956b). Biot used Lagrange's equations to derive a set of differential equations that govern the separated motions of a porous solid and a compressible viscous fluid confined to it. In 1962 Biot extended the acoustic propagation theory in a wider context of the dynamics of anisotropic media (Biot,1962a, 1962b). The theory is applied to the materials where fluid and solid are of comparable densities. As follows from Table 1, the densities of the composites differ from that of the polypropylene no more than 10% and the criterion of the applicability of the Biot theory is fulfilled. The plots of frequency dependence of sound absorption were derived by Biot for different combinations of elastic constants and densities of the porous solid and that of fluid taking into account the additional apparent mass due to inertia coupling. The theoretical curves (Biot, 1956b) exhibit a maximum value of the absorption at a characteristic frequency which depends on the kinematic viscosity of the fluid and pore diameter. The maxima are very pronounced in the case of fluid - saturated porous solids characterized by the elastic properties and densities far from the "compatibility condition":

$$z_1 = \frac{V_1^2}{V_c^2} \cong 1 ,$$

(5)

where V_1 stands for the velocity of the stress wave in a real anisotropic solid and V_c represents this velocity when the relative motion between fluid and solid is prevented in some way. The less the fraction z_1 the more enhanced are the maxima.

Acoustic standing wave method (Markiewicz et al., 2009) is the most popular way to determine the sound absorptive power of the material sample subjected to the plane acoustic wave. In this method, plane acoustic waves are generated by a loudspeaker placed at one end of a tube while the other end is terminated by the material sample. Due to the reflections from the sample, standing wave is produced in the tube as the superposition of the incident wave with the amplitude A and the reflected one with the amplitude B. The reflected wave is characterized by lower amplitude and shifted phase in comparison to the incident one. The probe microphone, moved inside the tube, receives the alternating acoustic pressure of minimum amplitude $p_{min}=A-B$ at the distance of $\lambda/4$ (λ – wavelength) from the sample followed by the pressure of maximum value $p_{max}=A+B$ at $\lambda/2$. As the acoustic wave energy is proportional to the square of the sound pressure, the equation (3) can be written:

$$\alpha = \frac{E_i - E_r}{E_i} = 1 - \left[\frac{B}{A}\right]^2 = 1 - \left[\frac{p_{max} - p_{min}}{p_{max} + p_{min}}\right]^2 ,$$ (6)

where E_r denotes the energy of the reflected wave. The equation (6) shows that the absorption coefficient α can be easily determined by means of the measurement of p_{min} and p_{max} amplitudes of the sound pressure inside the tube.

Figs. 1 and 2 show the results of the measurement of sound absorption coefficient α for pure polypropylene and the composites with lignocellulosic materials. The values of coefficient α were measured at the frequencies: 1000, 1800, 3000, 4000, 5000 and 6300 Hz according to the method mentioned above. As follows from the figures, the polypropylene is characterized by the relatively weak sound absorption. The values of coefficient α vary between 2 and 13% with the tendency to slightly decrease with increasing frequency. Fig. 1 shows the effect of the addition of 40 wt. % of hemp fillers on the absorption spectrum. The effect is dominant in the range of higher frequencies. Starting from the frequency of 3000Hz, the value of the coefficient α increases rapidly up to about 25% and maintains at this level up to 6300 Hz. Below the critical frequency 3000 Hz the effect of the addition of the hemp fillers is inconsiderable. One can even notice a small decrease of the coefficient α at 1800 Hz. Taking into account the fact that two different methods were used to prepare the samples No. 2 and No. 3 (extrusion and hydraulic pressing), one can conclude that the manufacturing procedure does not influence the sound absorption in the case of the composites with hemp filler. The composites containing long flax fibres, crumble rapeseed and crumble beech wood exhibit also better sound absorption in comparison with pure polypropylene but the frequency dependence of the coefficient α is quite different (Fig. 2). For these materials the maxima of α coefficient are observed at the frequencies of 3000 Hz or 4000 Hz. The differences in sound absorption by composites containing hemp and the ones based on fillers: flax, rapeseed straw, beech wood can be explained taking into account the Biot theory. The composites with hemp fillers seem to be nearest the "compatibility condition" among the investigated materials. The maxima of the absorption are not noticeable in the frequency range of the measurement, on the contrary to the remaining fillers. The composition of polypropylene and hemp results in such a combination of elastic constants and densities that the relative motion between filler and matrix is prevented. The discrepancies in sound absorption characteristics can result also from the filler morphology and its chemical composition. The width of hemp fibres (30 μm) is larger in comparison with flax fibres (20 μm). Moreover, hemp plant is known to have the dimensions of the anatomic cells larger than the remaining plants under examination. Hemp is distinguished for the highest contents of cellulose (75 wt. %) (Averous & Le Digabel, 2006). Flax contains 71 wt. % of cellulose. Beech and rapeseed are characterized by smaller contents of cellulose: beech – 42 wt. % (Nik–Azar et al., 1997) and rapeseed – from 35 to 40 wt. % (Paukszta, 2005, 2006). The contents of lignin in hemp (4 wt. %) is twice that of flax (2 wt. %) (Averous & Le Digabel, 2006). However, beech and rapeseed are known to have relatively large amount of lignin (~20 wt. %) (Paukszta, 2006). Flax is characterized by the contents of pectins (2 wt. %), fats (2 wt. %) and waxes (2 wt. %) which are twice those of hemp. It can be concluded that the higher contents of the cellulose in the hemp probably enables the sound absorption in the relatively wide frequency range.

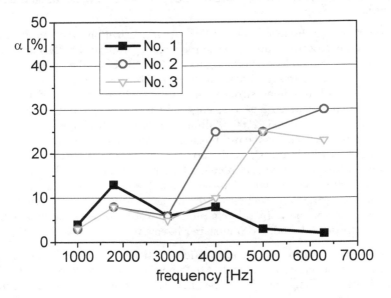

Fig. 1. Frequency dependences of sound absorption coefficient α for the samples: No. 1 - PP, No. 2 - PP + 40 wt. % of crumble hemp plant, No. 3 – PP + 40 wt. % of long hemp fibres

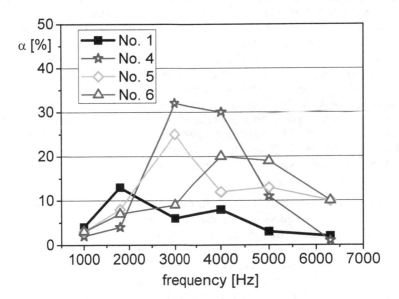

Fig. 2. Frequency dependences of sound absorption coefficient α for the samples: No. 1 – PP, No. 4 - PP + 40 wt. % of long flax fibres, No. 5 - PP + 40 wt. % of crumble rapeseed, No. 6 – PP + 40 wt. % of crumble beech wood

3.2 Effect of chemical treatment of lignocellulosic fillers on the acoustic properties of the composites

Chemical treatment of lignocellulosic filler as mercerization and modification with anhydride is often necessary to improve the mechanical properties of the composite due to better adhesion between the hydrophilic lignocellulosic filler and the hydrophobic polymer matrix (Borysiak & Garbarczyk, 2003; Borysiak & Doczekalska, 2005; Liu & Hu, 2008). Chemical treatment of the filler surface results in positive changes of the mechanical parameters of the composite as tensile strength, flexural strength and elongation at break (Bledzki et al., 2005; S.J. Kim et al., 2008; Mahlberg et al., 2001; Nachtigall et al., 2007; Yang et al., 2006). The interfacial region is the most vulnerable location to mechanical fracture. When subjected to the stress, it should show the ability to transmit the acoustic wave – induced tension from one phase to the other. Chemical treatment of the lignocellulosic filler is often necessary to get better the mechanical properties of the composite. Thus, the information about the influence of the modification on the acoustic properties is also very useful from the application point of view. In this chapter, the results of acoustic measurements for composites with crude and modified fillers are presented.

Figs. 3 – 6 show the values of sound absorption coefficient α measured for the composite samples No. 7 – 14, specified in Table 1. All composite samples, crude and modified, are compared with the pure polypropylene. The effect of the addition of the lignocellulosic filler to the polypropylene matrix is predominant in the frequency range above 3000 Hz where one can observe the improvement in the sound absorption of the order of more than ten percentage. All the investigated composite samples show the resonance characteristics of sound absorption with the maximum in the frequency range from 4000 to 6000 Hz. This behaviour can be ascribed to the combination of elastic constants and densities of both materials which makes possible the comparatively big relative motion between matrix and filler according to the Biot theory. The elastic constants and densities of the filler material result from its chemical composition. Our previous investigations, described in chapter 2.1, performed on composites with flax, hemp, rapeseed and beech fillers, showed that the increased sound absorption in a narrow frequency range can be probably related to the relatively small content of the cellulose. Now, we can confirm this presumption. The lignocellulosic materials used in this experiment are characterized by lower contents of cellulose: beech – 42 wt. % (Nik–Azar et al., 1997), pine – 41 wt. % (Gosselink et al., 2004), rapeseed Kaszub – 38 wt. % (Paukszta, 2005, 2006) and rape Californium - 37 wt. % (Paukszta, 2006) in comparison with hemp and flax which contain above 70 wt. % of cellulose (Averous & Le Digabel, 2006).

The effect of the modification of the fibre surface consists in the shift of the sound absorption maximum towards higher frequency range and is accompanied by the decrease of the sound absorption coefficient α. The reduction of the coefficient α is not large. It amounts 2.5 % in the case of composites with pine wood and rapeseed straw Californium. For the samples with beech wood filler the coefficient α remains unaffected after modification. The exception is the composite containing the rapeseed straw Kaszub as a filler that shows the decreasing in the sound absorption of 7% The reduction of the α coefficient value at the frequency related to the maximum of the sound absorption can be associated with the increase in the density of the composite after modification (Table 1) due to better adhesion between filler particles and polypropylene matrix. The fact implies that the specific acoustic impedance of

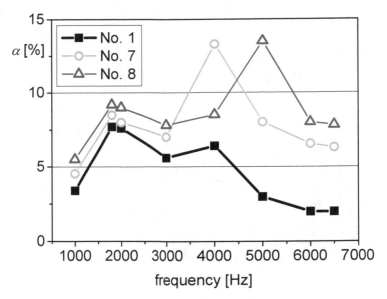

Fig. 3. Frequency dependences of sound absorption coefficient α for the samples: No. 1 – PP, No. 7 – PP + 20 wt. % of crude beech, No. 8 - PP + 20 wt. % of beech modified with succinic anhydride

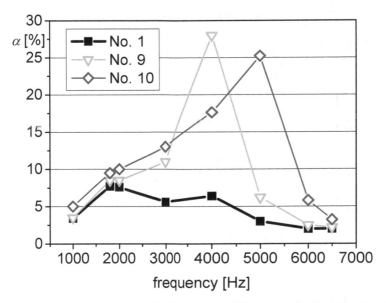

Fig. 4. Frequency dependences of sound absorption coefficient α for the samples: No. 1 – PP, No. 9 – PP + 20 wt. % of crude pine, No. 10 - PP + 20 wt. % of pine modified with succinic anhydride

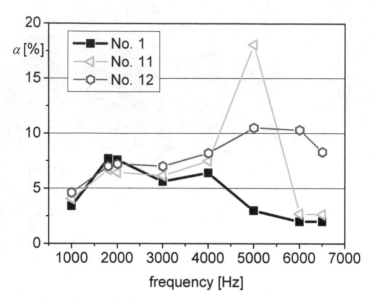

Fig. 5. Frequency dependences of sound absorption coefficient α for the samples: No. 1 – PP, No. 11 – PP + 30 wt. % of crude rapeseed straw Kaszub, No. 12 - PP + 30 wt. % of rapeseed straw Kaszub modified with acetic anhydride

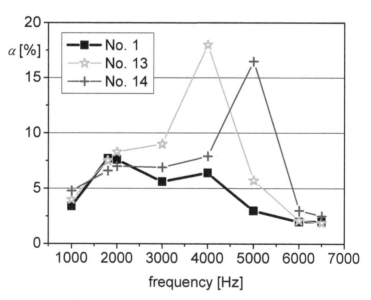

Fig. 6. Frequency dependences of sound absorption coefficient α for the samples: No. 1 – PP, No. 13 – PP + 30 wt. % of crude rapeseed straw Californium pure polypropylene (No. 1) and polypropylene composites with 30 wt. % of crude rapeseed straw Californium, No. 14 – PP + 30 wt. % of rapeseed straw Californium modified with acetic anhydride

the material, defined as the product of the density and the sound velocity (Lee & Chen, 2001), is increased. In turn, the higher the acoustic impedance the more acoustic energy is reflected from the material surface and the less of it can be absorbed. Better adhesion between polymer matrix and lignocellulosic filler leads to the better ability to transmit the acoustic wave – induced tension from one phase to the other. The process of the transmission in the composites with the modified fillers can be more rapid in comparison to that observed in non-modified composites and it can follow with higher frequency being a reason of a shift of the sound absorption band of about 1000 Hz.

4. Dielectric properties of the polypropylene reinforced with lignocellulosic materials

4.1 Effect of reinforcement of polypropylene with lignocellulosic materials on the dielectric properties

The dielectric properties of a material are determined by the polarizability of its molecules. There are three primary contributions to the electric polarization of a dielectrics: electronic, ionic and dipole reorientation – related (Uchino, 2000). The intensity with which each mechanism occurs depends on the frequency of applied electric field. The electronic polarization causes a displacement of the electrons with respect to the atomic nuclei and can follow alternating field with the frequencies up to $10^{12} - 10^{15}$ Hz. The ionic polarization relies on a displacement of the atomic nuclei relative to one another and responds up to $10^9 - 10^{12}$ Hz. Both mentioned polarization mechanisms are related to the non-polar molecules. The third mechanism associated with the dipole reorientation is valid only in the case of polar molecules. It can follow with the frequency of alternating electric field up to $10^6 - 10^9$ Hz. The dielectric permittivity ε' of a material represents the ratio of the capacitance of a plane condenser filled with the dielectric to that of the same condenser under vacuum and is to calculate from the expression:

$$\varepsilon' = \frac{C \cdot d}{\varepsilon_0 \cdot S} , \tag{7}$$

where: C is the capacitance of the condenser with the dielectric, S stands for the area of the sample covered by the electrode, d relates to the thickness of the sample and $\varepsilon_0 = 8.85 \cdot 10^{-12}$ F/m is the dielectric constant of the vacuum. The alternating current conductivity $\sigma_{a.c.}$ is described by the relationship:

$$\sigma_{a.c.} = \varepsilon_0 \cdot \omega \cdot \varepsilon' \cdot tg\delta , \tag{8}$$

where ω stands for the angular frequency.

The frequency dependencies of the dielectric constant ε' measured at room temperature for the polypropylene and its composites with various lignocellulosic materials derived from hemp and flax are presented in Figs. 7 and 8. The effect of the reinforcement of the polypropylene with a lignocellulosic material consists in the increase of the dielectric permittivity ε' over the whole measurement frequency range. The effect is predominant at lower frequencies. Pure polypropylene is a non-polar hydrophobic material which shows only instantaneous ionic and electronic polarization. Its dielectric permittivity ε' holds

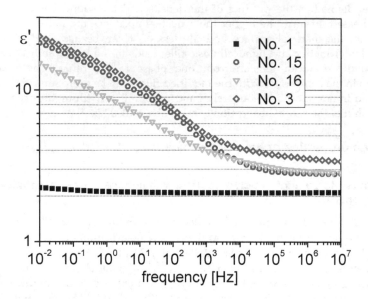

Fig. 7. Frequency dependences of dielectric permittivity ε' for the samples: No. 1 – PP, No. 15 - PP+25 wt. % of short hemp fibres, No. 16 - PP + 25 wt. % of hemp shivers, No.3 – PP + 40 wt. % of long hemp

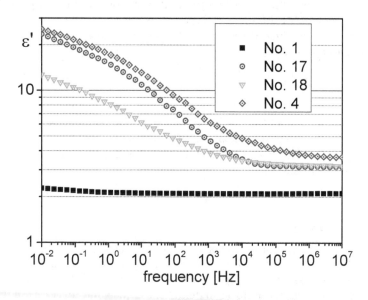

Fig. 8. Frequency dependences of dielectric permittivity ε' for the samples: No. 1 – PP, No. 17 – PP + 25 wt. % of short flax fibres, No. 18. – PP + 25 wt. % of flax shivers, No. 4 - PP + 40 wt. % of long flax fibres

nearly constant value in the whole frequency range with a slender increase bellow 10^3 Hz. The addition of a hydrophilic lignocellulosic material to the polypropylene entails the insertion of polar groups into the non-polar material giving the reason for rising the polarization related to the dipole reorientation. Moreover, the presence of the hydroxyl groups –OH in the cellulose, the hemicellulose and the lignin extends the moisture absorption due to the interaction of –OH groups and water molecules. The overall polarization of the composite, being the sum of three contributions: electronic, ionic and dipole reorientation – related ones, exhibits the maximum values at low frequencies and decreases with increasing frequency. The same behaviour shows the dielectric permittivity of the composites ε'. The value of the dielectric permittivity ε' increases with the content of the lignocellulosic material. In the higher frequency range, i.e. above 10^6 Hz, the value of the relative dielectric permittivity ε' tends to the constant value fixed by squared refractive index.

Moreover, the values for composites with the same content (25 wt.%) of different lignocellulosic materials converge. The differences in the dielectric permittivity ε' values are most significant in the low frequency range, and they are observed also for the composites comprising the same content of lignocellulosic materials derived from different parts of hemp and flax. The composites containing the shivers derived from hemp as well as from flax (25 wt.%) exhibit lower ε' than the ones comprising short fibres (also 25 wt.%). The shivers are the lignified parts of the stems, separated from the fibres and they show lower capacity of moisture absorption. This fact can indicate a smaller number of polar groups and lower polarization related to dipole reorientation.

Plots of the reciprocal of dielectric permittivity $1/\varepsilon'$ versus volume fraction of lignocellulosic material derived from hemp and flax at the frequency of 1 MHz (Fig. 9) are linear for the

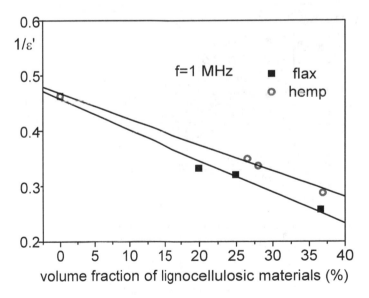

Fig. 9. Reciprocal dielectric permittivity $1/\varepsilon'$ versus volume fraction of lignocellulosic material derived from hemp and flax (Markiewicz et al., 2009)

applied contents of filler. A number of numerical relations as Lichtenecker, Maxwell Garnet, Jayasundere, Poon-Shin equations or Effective Medium Theory were developed by researchers to predict the effective dielectric constant of the composites (Subodh et al., 2007). All the mentioned models differ at higher volume fraction of the filler and they can be replaced by the linear fit at lower filler content, when the permittivity contrast between matrix and filler is low, particularly. For our investigated samples with the volume fraction less than 0.4 the dependence of the reciprocal of the dielectric permittivity on the volume fraction can be approximated by the linear fit (Sareni et al., 1997). The obtained results are in agreement with those presented by Jacob M. at al. (Jacob, 2006) for sisal-oil palm hybrid biofibre reinforced natural rubber biocomposites.

The reinforcement of the polypropylene with the lignocellulosic material results in the increase of ac conductivity (Figs. 10 and 11). The random distribution of the lignocellulosic fillers in the polypropylene matrix enables rearrangement of the fibres in a chain structure which ensures better carrier mobility in the presence of electric field. The frequency dependence of the electrical conductivity is described by the expression (Jonsher, 1997):

$$\sigma(\omega) \propto \omega^n . \tag{9}$$

The exponent n is close to 0.5 for pure polypropylene and points to diffusive carrier transport. For the composites, n changes from ~ about 0.5 at the low frequencies to ~ 1 at the high frequencies. This fact proves the existence of diffusive as well as hopping carrier transport. In the lowest frequency range the composite samples show the frequency-independent behavior pointing to the ohmic conduction. This property makes them to be better antistatic material than pure propylene.

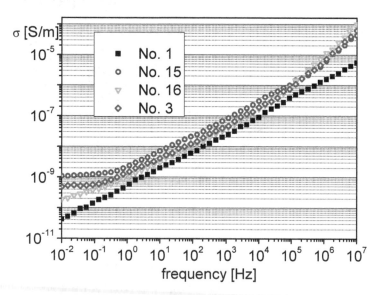

Fig. 10. Frequency dependences of ac conductivity σ' for the samples: No. 1 – PP, No. 15 – PP + 25 wt. % of short hemp fibres, No. 16 – PP + 25 wt. % of hemp shivers, No. 3 - PP+40% of long hemp fibres

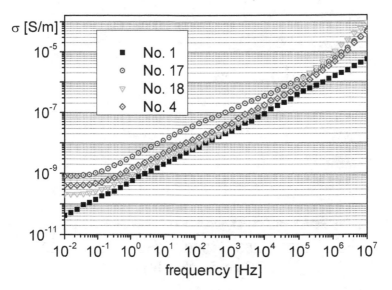

Fig. 11. Frequency dependences of ac conductivity σ for the samples: No. 1 – PP, No. 17 – PP + 25 wt, % of short flax fibres, No. 18 - PP+25 wt. % of flax shivers, No. 4 – PP + 40 wt. % of long flax fibres

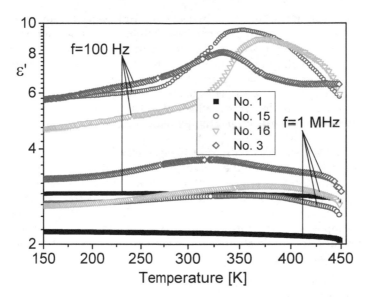

Fig. 12. Temperature dependences of dielectric permittivity ε' obtained at frequencies 100 kHz and 1 MHz for the samples: No. 1- polypropylene PP; No. 15 - PP + 25 wt.% of short hemp fibres; No. 16 – PP + 25 wt.% of hemp shivers; No. 3 – PP + 40 wt.% of long hemp fibres

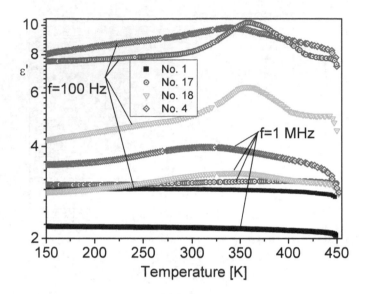

Fig. 13. Temperature dependences of dielectric permittivity ε' obtained at frequencies 100 kHz and 1 MHz for the samples: No. 1 - polypropylene PP; No. 17 - PP + 25 wt.% of short flax fibres; No. 18 - PP + 25 wt.% of flax shivers; No. 4 - PP + 40 wt.% of long flax fibres

The temperature variations of the dielectric permittivity ε' investigated for the polypropylene as well as the composite samples with hemp and flax are presented in Figs. 12 and 13. The value of ε' measured for polypropylene is nearly independent on the temperature up to the melting point at 438 K (Doh, 2005). The dielectric permittivity ε' of the composites increases with the temperature up to the maximum associated with the traces of water, and then decreases. The position of the maximum is determined by the contents of chemically bounded water which cannot be removed during the preparation. The maximum is shifted towards higher temperatures in the case of higher contents (Chand, 2005). As follows from Figs. 12 and 13, the technique of hydraulic pressing, applied for fabrication of composite samples with long fibers, implied the lowest content of water. In the vicinity of the melting point of the polypropylene, a rapid fall of the ε' value is visible. The dielectric permittivity ε' decreases with the increase of the frequency, as is seen for two frequencies: 100 Hz and 1 MHz. A weak dependency of dielectric permittivity ε' on the temperature was observed for the frequency of 1 MHz, particularly in the case of polypropylene composites with short fibres. This feature is the evidence that the composites can be recommended for application in the high frequency range because of the stable dielectric permittivity ε' value.

The dielectric loss factors ε'' of the pure polypropylene and the composites containing the lignocellulosic materials derived from hemp and flax are presented in Figs. 14 and 15 as a function of the temperature for the frequency of 1000 Hz. Pure polypropylene is known to exhibit two characteristic features (Kotek et al., 2005): a glass relaxation peak around 263 K and a high – temperature (~323 K) shoulder associated with chain relaxation in the crystalline phase. These features cannot be detected by the Dielectric Relaxation

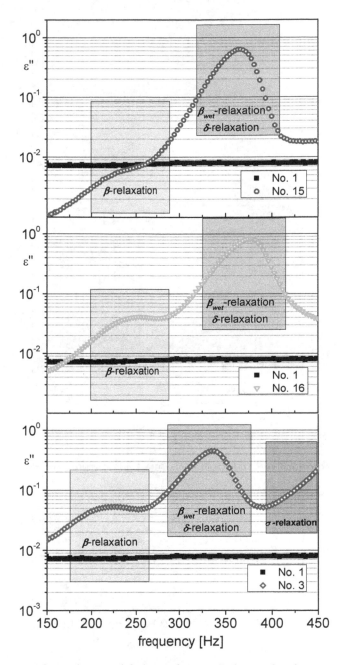

Fig. 14. Temperature dependences of dielectric losses ε'' obtained at frequency 1000 Hz for the samples: No. 1- polypropylene PP; No. 15 - PP + 25 wt.% of short hemp fibres; No. 16 – PP + 25 wt.% of hemp shivers; No. 3 – PP + 40 wt.% of long hemp fibres

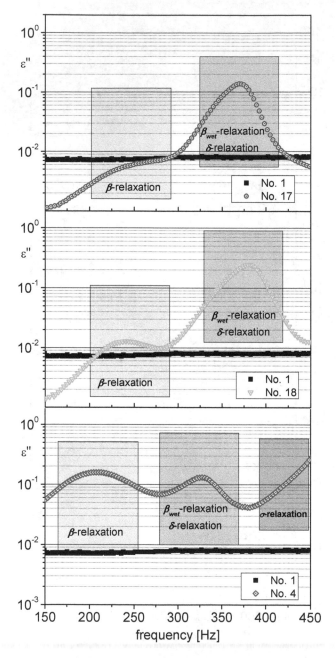

Fig. 15. Temperature dependences of dielectric losses ε'' obtained at frequency 1000 Hz for the samples: No. 1- polypropylene PP; No. 17 - PP + 25 wt.% of short flax fibres; No. 18 – PP + 25 wt.% of flax shivers; No. 3 – PP + 40 wt.% of long flax fibres

Spectroscopy (DRS) method without special modification of the polymer structure or introducing polar groups in the structure because polypropylene has no appreciable molecular dipoles and is not dielectrically active. The non-polar polypropylene does not show any anomalies in the dielectric loss spectrum. However, the temperature dependences of the dielectric loss factors ε'' measured for the composites are strongly influenced by the contribution of the lignocellulosic materials. Both kinds of the filler modify the dielectric absorption spectrum in the same way. In the low temperature range (from about 200 to 270 K) one can observe the maxima of ε'' ascribed to the β- relaxation process in the cellulose which is the main component of each lignocellulosic material. The β- relaxation is interpreted as a local motion of chain segments via the glucosidic linkages (Einfeldt et al., 2001). Above room temperature (from about 300 to 400 K) one can notice high relaxation peaks. Based on the shape of these peaks, one can deduce that two relaxation processes overlap in this temperature range: β_{wet} – the relaxation associated with the orientational motion of both cellulose and water (Baranov et al., 2003; Einfeldt et al., 2001) and δ- the relaxation ascribed to the motion of the end groups in branched polymers (Einfeldt et al, 2001) present in the lignocellulosic material (hemicellulose, pectin, lignin). Because the intensity of the δ-relaxation is significantly smaller than that of β_{wet} - relaxation (Einfeldt et al., 2001), one can state that in the polypropylene – lignocellulosic materials composites the β_{wet}- relaxation is disturbed by the δ- process. In the case of the composites with long fibres derived from flax as well as hemp, the increase in the dielectric losses was observed in the highest temperature range (above 420 K). The effect results from the electric conductivity and is called σ- relaxation (Einfeldt et al., 2001). The losses due to the electric conductivity are ascribed to charge carrier hopping between localized sites in amorphous solids. The fact that the σ - relaxation was observed only for the composites with long fibres confirms the conclusion from (Einfeldt et al., 2001) that the activation energy for the carrier hopping increases when the amount of water is reduced. The intensity of β_{wet} – relaxation is proportional to the contents of water. The low intensity of β_{wet} – relaxation in the composites with long fibers is a reason for relatively high strength of β – relaxation in comparison with that observed for other investigated samples where the β – process is suppressed by the β_{wet} – relaxation and the position of high intensive β_{wet} – relaxation peak in higher temperature range masks the σ- process.

4.2 Effect of chemical treatment of lignocellulosic fillers on the dielectric properties of the composites

The effect of chemical treatment is dominant in the low frequency range, i. e. from 10^{-2} Hz to 1 kHz. It can be opposite for various kinds of lignocellulosic fillers. Figs. 16 and 17 show the frequency dependences of dielectric permittivity ε' obtained at room temperature for the polypropylene composites containing crude, mercerized and modified lignocellulosic fillers derived from pine and beech wood as well as two kinds of rapeseed straw. The effect observed for the pine and beech wood (Fig. 16) consists in the increase of the dielectric permittivity ε' value. The modification with maleic anhydride causes greater increase than the mercerization. In the case of both kinds of rapeseed straw: Kaszub and Californium (Fig. 17), the mercerization decreases the dielectric permittivity ε' value several times and the modification with maleic anhydride reduces ε' value to that measured for composites with crude pine and crude beech. One should take into account the location of the β_{wet}- relaxation in the vicinity of room temperature to explain the opposite influence of chemical treatment

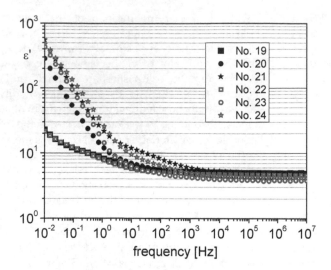

Fig. 16. Frequency dependences of dielectric permittivity ε' obtained for the samples: No. 19 – PP + 30 wt. % of crude beech; No. 20 – PP + 30 wt. % of mercerized beech; No. 21 – PP + 30 wt. % of beech modified with maleic anhydride; No. 22 – PP + 30 wt. % of crude pine; No. 23 – PP + 30 wt. % of mercerized pine; No. 24 – PP + wt. 30% of pine modified with maleic anhydride

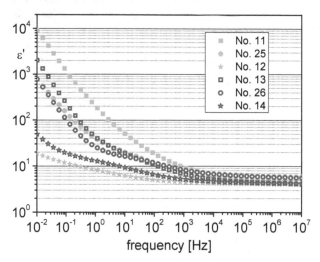

Fig. 17. Frequency dependences of dielectric permittivity ε' obtained for the samples: No. 11 – PP + 30 wt. % of crude rapeseed straw Kaszub; No. 25 – PP + 30 wt. % of mercerized rapeseed straw Kaszub; No. 12 – PP + 30 wt. % of rapeseed straw Kaszub modified with acetic anhydride; No. 13 – PP + 30 wt. % of crude rapeseed straw Californium; No. 26 – PP + 30 wt. % of mercerized rapeseed straw Californium; No. 14 – PP + 30 wt. % of rapeseed straw Californium modified with acetic anhydride

on the dielectric permittivity ε' value at low frequencies. The effect is associated with the ability of moisture absorption. The swollen structure of crude rapeseed straw facilitates the moisture absorption and the ε' value for this material is the biggest one. The mercerization and the modification with acetic anhydride make the structure more rigid and the moisture absorption is limited. However, in the case of beech and pine the crystalline structure of the cellulose confined to the fibres gets be swelled just after chemical treatment.

5. Conclusion

The investigated composite materials based on polypropylene matrix and lignocellulosic fillers can be recommended for application in building and automotive industry because of their good sound absorptive power as well as in packaging electronics due to their dielectric properties. Addition of lignocellulosic materials to the pure polypropylene increases the sound absorption coefficient by about 20% in the frequency range above 3000 Hz. The frequency dependence of the absorption coefficient can be shaped by the proper choice of the lignocellulosic filler. The shift of the sound absorption band due to chemical treatment of the lignocellulosic fibres seems to be suitable for manufacturing the composites with extended sound absorption frequency range as the effect of the adaptation of mixed filler containing the treated and untreated lignocellulosic material. The increased dielectric permittivity of the composites, in comparison with the polypropylene, is preferable in the application in the field of packaging industry. The polypropylene-lignocellulosic materials composites assure the thermal stability of the dielectric permittivity above 1 MHz and better antistatic properties than pure polypropylene.

6. Acknowledgment

The authors are grateful to Professor Józef Garbarczyk for inspiration and valuable discussion of the material presented in this manuscript. This research was supported by Grant of Poznan University of Technology 32-171/12-DS.

7. References

Averous, L. & Le Digabel, F. (2006). Properties of Biocomposites Based on Lignocellulosic Fillers. *Carbohydrate Polymers,* Vol.66, No.4, (November 2006), pp. 480-493, ISSN 0144-8617

Baranov, A.I.; Anisimova, V.N.; Khripunov, A.K. & Baklagina Y.G. (2003). Dielectric Properties and Dipole Glass Transition in Cellulose Acetobacter Xylinium. *Ferroelectrics,* Vol.286, No.1, (n. d. 2003), pp. 141-151, ISSN 0015-0193

Biot, M. A. (1955). Theory of Elasticity and Consolidation for a Porous Anisotropic Solid. *Journal of Applied Physics,* Vol.26, No.2, (February 1955), pp. 182-185, ISSN 0021-8979

Biot, M. A. (1956a). Theory of Propagation of Elastic Waves in a Fluid-Saturated Porous Solid. I. Low-Frequency Range. *Journal of the Acoustical Society of America,* Vol.28, No.2, (March 1956), pp. 168-178, ISSN 0001-4966

Biot, M. A. (1956b). Theory of Propagation of Elastic Waves in a Fluid-Saturated Porous Solid. II. Higher Frequency Range. *Journal of the Acoustical Society of America,* Vol.28, No.2, (March 1956), pp. 179-191, ISSN 0001-4966

Biot, M.A. (1962a). Mechanics of deformation and acoustic propagation in porous media. *Journal of Applied Physics*, Vol.33, No.4, (April 1962), pp. 1482-1498, ISSN 0021-8979

Biot, M.A. (1962b). Generalized theory of acoustic propagation in porous dissipative media. *Journal of the Acoustical Society of America*, Vol.34, No.9A, (September 1962), pp. 1254-1264, ISSN 0001-4966

Bledzki, A.K. & Gassan, J. (1999). Composites Reinforced with Cellulose Based Fibres. *Progress in Polymer Science*, Vol.24, No.2, (May 1999), pp. 221-274, ISSN 0079-6700

Bledzki, A.K.; Letman, M.; Viksne, A. & Rence, L. (2005). A Comparison of Compounding Processes and Wood Type for Wood Fibre−PP Composites. *Composites: Part A*, Vol.36, No.6, (June 2005), pp. 789-797, ISSN 1359-835X

Borysiak, S. & Garbarczyk, J. (2003). Crystallisation of Isotactic Polypropylene with β-Nucleating Agents under Elevated Pressure. *Fibres and Textiles in Eastern Europe*, Vol.11, No.5, (January/December 2003), pp. 50-53

Borysiak, S.; Paukszta, D. & Helwig, M. (2006). Flammability of Wood-Polypropylene Composites. *Polymer Degradation and Stability*, Vol.91, No.12, (December 2006), pp. 3339-3343, ISSN 0141-3910

Borysiak, S. & Doczekalska, B. (2005). X-Ray Diffraction Study of Pine Wood Treated with NaOH. *Fibres and Textiles in Eastern Europe*, Vol.13, No.5, (January/December 2005), pp. 87-89

Chand, N. & Jain, D. (2005). Effect of Sisal Fibre Orientation on Electrical Properties of Sisal Fibre Reinforced Epoxy Composites. *Composites: Part A*, Vol.36, No.5, (May 2005), pp. 594-602, ISSN 1359-835X

Chung , D.L. (1995). *Materials for electronic packaging*, Butterworth–Heinemann, ISBN 0-7506-9314-2, Boston, USA

Doh G.H.; Lee, S.Y.; Kang, I.A. & Kong Y.T. (2005). Thermal Behavior of Liquefied Wood Polymer Composites. *Composite Structures*, Vol.68, No.1, (April 2005), pp. 103-108, ISSN 0263-8223

Einfeldt, J.; Meissner, D. & Kwasniewski, A. (2001). Polymerdynamics of Cellulose and Other Polysaccharides in Solid State-Secondary Dielectric Relaxation Processes. *Progress in Polymer Science*, Vol.26, No.9, (November 2001), pp. 1419-1472, ISSN 0079-6700

Epstein, P.S. & Carhart, R.R. The Absorption of Sound in Suspensions and Emulsions. I. Water Fog in Air. *Journal of the Acoustical Society of America*, Vol.25, No.3, (May 1953), pp. 553-565, ISSN 0001-4966

Ersoy, S. & Kucuk, H. (2009). Investigation of industrial tea-leaf-fibre waste materialfor its sound absorption properties. *Applied Acoustics*, Vol.70, No.1, (January 2009), pp. 215-220, ISSN 0003-682X

Gosselink, R.J.A.; Krosse, A.M.A.; van der Putten, J.C.; van der Kolk, J.C.; de Klerk–Engels, B. & van Dam, J.E.G. (2004). Wood Preservation by Low-temperature Carbonisation. *Industrial Crops and Products*, Vol.19, No.1, (January 2004), pp. 3-12, ISSN 0926-6690

Hattotuwa, G.B. Premalal; Ismail, H. & Baharin, A. (2002). Comparison of the Mechanical Properties of Rice Husk Powder Filled Polypropylene Composites with Talc Filled Polypropylene Composites. *Polymer Testing*, Vol.21, No.7, (n d 2002), pp. 833-839, ISSN 0142-9418

Jacob, M.; Varughese, K.T. & Thomas, S., (2006). Dielectric Characteristics of Sisal–Oil Palm Hybrid Biofibre Reinforced Natural Rubber Biocomposites. *Journal of Materials Science*, Vol. 41, No. 17, (September 2006), ISSN 0022-2461

Jonsher, A.K., (1977). The "Universal" Dielectric Response. *Nature*, Vol.267, No.5613, (June 1977), pp. 673-679, ISSN 0028-0836

Kim, S.J.; Moon, J.B.; Kim, G.H. & Ha, C.S. (2008). Mechanical Properties of Polypropylene/Natural Fiber Composites: Comparison of Wood Fiber and Cotton Fiber. *Polymer Testing*, Vol.27, No. 7, (October 2008), pp. 801-806, ISSN 0142-9418

Kotek, J.; Kelnar, I.; Studenovsky, M. & Baldrian J. (2005). Chlorosulfonated polypropylene: preparation and its application as a coupling agent in polypropylene–clay nanocomposites. *Polymer*, Vol. 46, No. 16, (June 2005), pp. 4876-4881, ISSN 0032-3861

Lee, F.C. & Chen, W.H. (2001). Acoustic Transmission Analysis of Multi-Layer absorbers. *Journal of Sound and Vibration*, Vol. 248, No.4, (December 2001), pp. 621-634, ISSN 0022-460X

Liu, Y. & Hu, H. (2008). X-ray Diffraction Study of Bamboo Fibers Treated with NaOH. *Fibers and Polymers*, Vol.9, No.6, (December 2008), pp. 735-739, ISSN 1229-9197

Mahlberg, R.; Paajanen, L.; Nurmi, A.; Kivisto, A.; Koskela, K. & Rowell, R.M. (2001). Effect of Chemical Modification of Wood on the Mechanical and Adhesion Properties of Wood Fiber/Polypropylene Fiber and Polypropylene/Veneer Composites. *Holz als Roh-und Werkstoff*, Vol.59, No.5, (October 2001), pp. 319-326, ISSN 0018-3768

Markiewicz, E.; Borysiak, S. & Paukszta; D. (2009). Polypropylene-lignocellulosic material composites as promising sound absorbing materials. *Polimery*, Vol.54, No.6, (June 2009), pp. 430-435, ISSN 0032-2725

Markiewicz, E.; Paukszta D. & Borysiak S., (2009). Dielectric Properties of LignocellulosicMaterials–Polypropylene composites. *Materials Science-Poland*, Vol.27, No.2, (n. d. 2009), pp. 581-593, ISSN 0137-1339

Mohanty, A.K.; Misra, M. & Hinrichsen, G. (2000). Biofibres, Biodegradable Polymers and Biocomposites: An Overview. *Macromolecular Materials and Engineering*, Vol.276-277, No. 1, (March 2000), pp. 1 -24, ISSN 1439-2054

Nachtigall, S.M.B.; Cerveira, G.S. & Rosa, S.M.L. (2007). New Polymeric-Coupling Agent for Polypropylene/Wood-Flour Composites. *Polymer Testing*, Vol.26, No.5, (August 2007), pp. 619-628, ISSN 0142-9418

Nik–Azar, M.; Hajaligol, M. R.; Sohrabi ,M. & Dabir, B. Mineral matter effects in rapid pyrolysis of beech wood . *Fuel Processing Technology*, Vol.51, No.1, (March 1997), pp. 7-17, ISSN 0378-3820

Nor, M.J.M.; Jamaluddin, M. & Tamiri, F.M. (2004). A Preliminary Study of Sound Absorption Using Multi-layer Coconut Coir Fibers. *Electronic Journal "Technical Acoustics"*, Vol.3, (March 2004), pp. 1-8, ISSN 1819-2408, Retrieved from http://ejta.org/en/tamiri1

Paukszta, D. Investigations of Lignocellulosic Materials from Rape for the Purpose of Producing Composites with Thermoplastic Polymers. *Fibres and Textiles in Eastern Europe*. Vol.13, No.5, (January/December 2005), pp. 90-92, ISSN 1230-3666

Paukszta, D. Chemical Composition of Wooden Parts of Rape Stem. *Oilseed Crops*, Vol. XXVII, No.1, (n. d. 2006), pp. 143-150

Pecht, M.G.; Agarwal, R.; McCluskey, P.; Dishonhg, T.; Javadpour, S.& Mahajan, R. (1999). *Electronic packaging materials and their properties*, CRC Press LLC, Washington, USA

Peijs, T. (2003). Composites for Recyclability. *Materials Today*, Vol.6, No.4, (April 2003), pp. 30-35, ISSN 1369-7021

Polish Patent 186577, 2004

Polish Patent 190405, 2005

Sareni, B.; Krähenbühl, L.; Beroual, A. & Brosseau, C. (1997). Effective Dielectric Constant of Random Composite Materials. *Journal of Applied Physics*, Vol.81, No.5, (March 1997), pp. 2375 – 2383 , ISSN 0021-8979

Soluch, W. (1980). *Wstęp do piezoelektroniki*, Wydawnictwa Komunukacji i Łączności, ISBN 83-206-0041-3, Warsaw, Poland

Subodh, G.; Pavithran, C.; Mohanan, P. & Sebastian, M.T. (2007). $PTFE/Sr_2Ce_2Ti_5O_{16}$ polymer ceramic composites for electronic packaging applications. *Journal of the European Ceramic Society*, Vol.27, No.8-9, (n. d. 2007), pp. 3039-3044, ISSN 0955-2219

Tiwari, V.; Shukla, A. & Bose A. (2004). Acoustic Properties of Cenosphere Reinforced Cement and Asphalt Concrete. *Applied Acoustics*, Vol.65, No.3, (March 2004), pp. 263 -275, ISSN 0003-682X

Tummala, R.R. (1991). Ceramic and Glass-Ceramic Packaging in the 1990s. *Journal of the American Ceramic Society*, Vol.74, No.1, (May 1991), pp. 895-908, ISSN 1551-2916

Uchino, K. (2000). *Ferroelectric Devices*, Marcel Dekker, ISBN 0-8247-8133-3, New York-Basel

Vinogradov, N. (2004). Physicochemical and Acoustic Properties of Water-Based Magnetic Colloid. *Colloid Journal*, Vol.66, No.11, (January 2004), pp.29-37, ISSN 1061-933X

Yang, H.S.; Kim, D.J.; Lee, Y.K. & Kim, H.J. (2003). Rice Straw–Wood Particle Composite for Sound Absorbing Wooden Construction Materials. *Bioresource Technology*, Vol.86, No.2, (January 2003), pp. 117-121, ISSN 0960-8524

Yang, H.S.; Wolcott, M.P.; Kim, H.S.; Kim, S.& Kim, H.J. (2006). Properties of Lignocellulosic Material Filled Polypropylene Biocomposites Made with Different Manufacturing Processes. *Polymer Testing*, Vol.25, No.5, (August 2006), pp. 668-676, ISSN 0142-9418

12

Effects on Freeze-Thaw Durability of Fibers in Concrete

Salih Taner Yildirim[1] and Cevdet Emin Ekinci[2]
[1]Kocaeli University, Civil Engineering Department, Kocaeli,
[2]Fırat University, Construction Education Department, Elazığ,
Turkey

1. Introduction

The investigation on a construction having such properties as durability, ductility, toughness and strength has boosted interest in materials, such as concrete with fiber and high performance (Otter & Naaman, 1988; Ramakrishnan et al., 1996; P.B. Cachim et al., 2002; Singh & Kaushik, 2003). While improvement of durability of the concrete depends on these conditions, contributory improvements in both chemical and mechanical properties of the concrete are also essential. The fibers improving concrete mechanically are primarily added to minimize cracking or to increase ductility of the concrete and fracture toughness against impact or dynamic loads (Naaman & Reinhardt, 1996; Dias&Thaumaturgo, 2005).

Since short fiber types greatly increase the number of fibers used in the concrete, they are used to decrease cracking and increase durability depending on the properties of the materials used; whereas, long fibers aim more often to increase mechanical properties of the concrete. Addition of hybrid fibers created synergy in the concrete and leaded to similar significant improvements in monofiber reinforced concrete having the higher total fiber content (Qian & Stroeven, 2000).

Studies on durability have emphasized that fiber gains particular importance on increasing freeze-thaw durability. Dramatic falls have been observed in elasticity modulus reaching as many as 300 cycles in the concrete specimens kept both in a 5 % NaSO$_4$ solution and in water, whose rate of w/c was 0.26, 0.32 and 0.44. Concrete including steel fibers do fail in much higher cycles than do plain concrete. Also, the concrete specimens, including fibers with the same rate of w/c, kept in the solution have been shown to have a higher performance in comparison with concrete without fibers kept in water (Singh & Kaushik, 2003).

In some studies, external loads have been exerted upon the concrete kept in a NaCl solution under the influence of effective freeze-thaw cycles. The concrete specimens exposed to NaCl have been shown to lose twice as much weight as those exposed to water. Specimens with steel fibers lose weight maximum at the w/c ratio of 0.44 that becomes obvious after 20-25 cycles. As the rate of the tension of the burdens exerted increased, resistance of the concrete specimens to cycles decreased. Addition of steel fibers in concrete specimens has been shown to cause a delay in a decrease in the performance of the concrete in the advanced cycles in comparison with the concrete without fiber (Sun et al., 2002; Mu et al., 2002; Miao et al., 2002). Morgan (1991) tested dry and wet hybrid shotcretes with sprayed fibers according to ASTM C 666 rapid water freeze-thaw method (A). In the case of air entraining and use of a high amount of steel and polypropylene fibers freeze-thaw durability can be achieved in wet and dry sprayed concrete. However, a rapid fall is observed in durability when there is no air entraining.

According to ASTM C 666 rapid water freeze-thaw method, 300 and 700 freeze-thaw cycles were applied. The results have shown that despite falls in durability and dynamic elasticity modulus, freeze-thaw performance of both concrete have been shown to be perfect. A study by Juska et al. (1999) on thermal effects upon concrete with glass fibers reported interesting results pertaining to durability. Gomez and Casto (1996) conducted the experiment of freeze (-17.8 °C) and thaw (4.4 °C) came in a 2 % NaCl solution upon composites with fibers. Flexural strength of the specimens and some other properties has been reported to suffer substantial loss. Moreover, Myers et al. (2001) reported flexural rigidity and strength in plaques with glass fibers to decrease more than did rigidity and strength in plaques with carbon fibers.

In this study, micro-structured polypropylene and glass fibers were both used separately and in combination with macro-structured steel fibers in the concrete. Experiments were conducted in order to determine weight-loss and durability factor based on ultrasound pulse velocity of 12 different concrete series produced according to ASTM C-666. The separate and combined effects of the fibers used in the concrete in terms of the rapid freeze-thaw period were investigated.

2. Experimental study

2.1 Materials

CEM I 42.5 R cement was used in the study (Yildirim, 2002). As the aggregate, crushed stone dust of 0-0.5 mm, natural sand of 0.5-4 mm, natural coarse aggregate of 4-16 mm and crushed stone of 16-32 mm were used. Densities of the aggregate used were 2.62, 2.65, 2.70, 2.70 gr/cm^3, respectively. The largest dimension of aggregate was 22 mm. Hooked steel fibers (SF), plain glass fiber (GF) and polypropylene fibers (PF) apart from additive materials providing superplasticizer was used. The properties of fibers have been presented in Table 1, while the properties of cement and additive providing super plasticizer have been presented in Table 2.

Properties	SF	PF	GF
Size (mm)	60	20	12
Dimension (mm)	0.75	0.05	0.014
Brittleness	80	400	857
Density (kg/mm^3)	7480	910	2680
Modulus of elasticity (MPa)	200000	3500-3900	72000
Tensile strength (MPa)	1100	320-400	1700
The number of fibers per kilogram	4600	82 Million	200 Million

Table 1. Technical properties of steel, polypropylene and glass fibers.

Concrete including fibers with different percentages were produced within the same main compounds. K represents the control concrete specimen, while S, P and G represent the concrete specimens including steel, polypropylene and glass fibers, respectively. Three series of concrete, including 0.5 (S 0.5), 0.75 (S 0.75) and 1 (S 1) % as volumetric of hooked steel fibers respectively, were produced. Another six series of concrete were produced with the use of polypropylene and glass fibers of 0.1 % (P and G). The symbols of the fiber material used in the mixture fiber specimens have been defined as SP and SG. They were both used separately and in combination with macro-structured steel fibers of 0.5 (SP 0.5 and SG 0.5), 0.75 (SP 0.75 and SG 0.75) and 1 (SP 1 and SG 1)%. Six specimens were produced from all the series so that they would be used in the experiments. Therefore, all of the twelve series of concrete were produced.

Chemical Compound	SiO_2	Al_2O_3	Fe_2O_3	CaO	MgO	SO_3	Na_2O	K_2O	Cl⁻	Insoluble Remains	LOI	C	S
Cement	20.03	5.09	3.44	64.5	1.43	1.3	0.19	0.65	0.01	0.42	2.31	-	-
Superplasticizer	90	0.75	1	0.65	1.15	-	0.55	1.05	0.03	-	-	1.90	0.23

Table 2. Chemical properties of cement and superplasticizer.

As to the control concrete compound of 1 m³, slump was constant as 130 mm. 181.0 kg of water, 281.9 kg of cement and 14.83 kg of silica fume were used, apart from superplasticizer. Water/cement rate was taken as 0.61. Fibers used as volumetric in concrete.

2.2 Experimental method

Freeze-thaw tests in the concrete series were conducted according to ASTM C 666 (Procedure B: rapid freeze-thaw under air conditions). It was applied on standard prismatic specimens at the dimensions of 80x80x360 mm. The experiment was conducted after the specimens had been cured for 28 days. Heat transfer calculations were made in order to determine how long the heat in the specimens would take to reach the optimal heat required for the experiment so that the freeze-thaw test could be made in the deep frost according to ASTM C 666. Therefore, the central heat of the specimens (20 °C) was lowered to -20 °C. The central heat of the specimens left in the water was increased to 5.4 °C. 30 cycles were made altogether, with the central heat adjusted in such a way that it would vary between -20 °C (2 hours and 40 minutes) and 5.4 °C (36 minutes).

Weight loss and ultrasound pulse velocity were measured at the beginning and the end of the cycles applied on concrete specimens. The percentage of dynamic E-modulus determined after freeze-thaw cycles through ultrasound instrument (P) was calculated by dividing the square of the pulse velocity after freeze-thaw cycles by the square of the pulse velocity before freeze-thaw cycles and then multiplying the result by 100. Afterwards, dynamic E-modulus was determined by multiplying a predetermined value (30) by the continuing number of cycles (N). The result was then divided by the final number of cycle (30) (M). After this, durability factor (DF) for the concrete specimens was determined with

DF=P.N/M formula. N and M values were equal for this study (American Society for Testing and Materials [ASTM C 666], 1999).

3. Test results and discussion

As a result of the experiment, surface damages of concrete specimens were determined after 30 freeze-thaw cycles. The specimens were seen that they have a sponge-like surface and broken edge. These damages occurred more intensively between 20 and 25 cycles (Mia et al., 2002).

As seen in Fig. 1, the plain concrete specimens had far less amounts of weight-loss than most of the concrete specimens including fiber. However, there occurred roughness over the surface. In order to obtain a more precise determination of weight-loss, the number of freeze-thaw cycles should be increased. The surface of concrete gets damaged when exposed to low freeze-thaw cycles. A much less amount of weight-loss could have been expected in fiber reinforced concrete if they had been exposed to a larger number of freeze-thaw cycles, considering the capability of fibers to keep matrixes together. Weight-loss was also determined to get affected by some of the parts falling off the corners of the specimens. In particular, the corners of the concrete with steel fibers are weaker and so may cause falling off some parts in the corners by forcing the matrix during the freeze-thaw cycles.

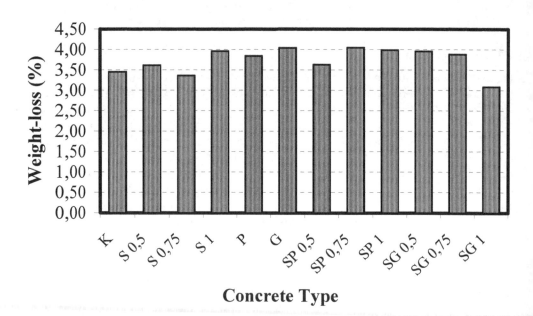

Fig. 1. Weight-loss after freeze-thaw cycles.

As seen in Fig. 2, the decrease in the pulse velocity of the specimens was obvious in contrast to their weight-loss. Polypropylene fibers demonstrated the best performance as a stand-alone and mixed with steel fibers (Morgan, 1991). A similar effect was determined for the mixture fibers. Fibers did affect the concrete specimens in agreement with their own properties. Because of the capability of polypropylene fibers to prevent cracking and to be remarkably safe from corrosion, the pulse velocity value was determined to be low.

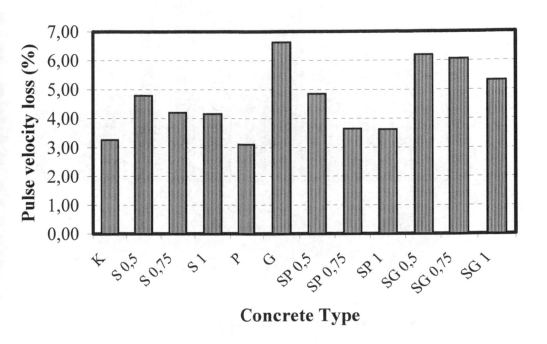

Fig. 2. Decreases in the pulse velocity following freeze-thaw cycles.

Though glass fibers have microstructures like polypropylene fibers, they showed their general weakness and the fibers were observed to have caused voids and to have affected adhesion negatively (Juska,1999). The initial decrease was in the steel fibers were attributed to the voids between the fiber-matrix interfaces. Moreover, due the fact that the ends of the fibers were curved, and that they are more resistant to dilation or contraction, the pulse velocity decreases due to the increase in the amount of fiber. However, it should be emphasized that the following cycles may change this. In other words, it is possible to say that the durability factor may be higher in following cycles (Singh & Kaushik, 2003). In order to ensure this, the number of cycles should over 30. Because the durability factors seen in Fig. 3 are inversely proportional to the decrease in the ultrasound pulse velocity, the low values are seen as high here.

Durability of the concrete series was determined to have been increased by polypropylene fibers, while glass fibers were determined to be highly unsuccessful. No concrete specimens including steel fibers, inclusive of mixture fibers, had higher values than plain concrete specimens.

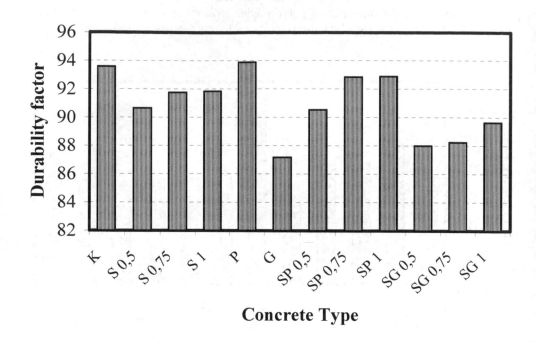

Fig. 3. Durability factors determined after freeze-thaw cycles.

4. Conclusion

20-25 of freeze-thaw cycles in concrete specimens was determined to be highly critical. Though they seem to be enough in number to form an opinion about the durability factor, 30 cycles of the freeze-thaw process seem to be insufficient to determine the precise weight loss of the concrete specimens. It was concluded that increasing the number of cycles could be advantageous to measuring weight, and that steel fibers could provide advantages for durability determination of in the following cycles. Using polypropylene fiber in all the concrete specimens to be reinforced by fibers in consideration of the effects of freeze-thaw cycles will be advantageous. Steel fibers did not cause any difference in terms of freeze-thaw cycles but did cause some negative effects for mixture concrete. Therefore, it is important that glass fibers should not be used in the places exposed to freeze-thaw cycles, or that concrete should be protected against this effect. Base on the study results, it can be suggested that steel fibers, which have different levels of brittleness, those with smaller dimensions in particular, should be investigated for the effects of freeze-thaw cycles.

5. References

ASTM C 666 (1999). Standard Test Method for Resistance of Concrete to Rapid Freezing and Thawing, Volume 04.02,
Concrete and Aggregates, *American Society for Testing and Materials*, West Conshohocken, PA.

Cachim, P.B.; Figueiras, J. A. & Pereira, P. A. A. (2002). Fatigue Behavior of Fiber-Reinforced Concrete in Compression, *Cement and Concrete Composites*, Vol. 24, pp.211–217, ISSN 0958-9465.

Otter, D.E., Naaman, A.E. (1988). Properties of Steel Fiber Reinforced Concrete Under Cyclic Loading, *ACI Materials J.*, Vol. 85, No. 4, pp. 254–261. ISSN 0889325X.

Dias, D.P., Thaumaturgo, C. (2005). Fracture toughness of geopolymeric concrete reinforced with basalt fibers, *Cement and Concrete Composites*, Vol. 27, pp. 49–54, ISSN 0958-9465.

Gomez, J.; Casto, B. (1996). Freeze-Thaw Durability of Composite Materials, *Proceedings of the 1st International Conference on Composites in Infrastructure, Fiber Composites in Infrastructure*, pp. 947-955.

Juska, T.; Dutta, P.; Carlson, L.; Weitsman, J.(1999). Gap Analysis for Durability of Fiber Reinforced Polymer Composites in Civil Infrastructure, *Thermal Effects*, Chapter 5, pp.40-51.

Miao, C.; Mu, R.; Tian, Q. & Sun, W. (2002). Effect of Sulphate Solution on the Frost Resistance of Concrete with and without Steel Fiber Reinforcement, *Cement and Concrete Research*, Vol. 32, No. 1, pp. 31–34, ISSN 0008-8846.

Morgan, D.R. (1991). Freeze Thaw Durability of Steel and Polypropylene Reinforced Shotcretes: a Review. Durability of Concrete. *Second International Conference*, held in Montreal, Canada; Ed. by V.M. Malhotra; American Concrete Institute, Detroit, MI, Vol. 2, pp. 901-918. (ACI SP-126).

Mu, R.; Miao, C.; Luo, X.; Sun, W. (2002). Interaction between Loading, Freeze-Thaw Cycles, and Chloride Salt Attack of Concrete with and without Steel Fiber Reinforcement, *Cement and Concrete Research*, Vol. 32, pp. 1061-1066, ISSN 0008-8846.

Myers, J.J.; Murthy, S.; Micelli, F. (2001). "Effect of Combined Environmental Cycles on the Bond of FRP Sheets to Concrete," Proceedings - Composites In Construction, *2001 International Conference*, Porto, Portugal, October 10-12.

Naaman, A. & Reinhardt, H.W., Eds (1996). High Performance Fiber Reinforced Cement Composites (HPFRCC2), *Proc., 2nd Int. RILEM Workshop*, RILEM, Proceedings 31, London, E.&FN Spon, ISBN 0419211802.

Qian, C.X. & Stroeven, P. (2000). Development of hybrid Polypropylene-Steel Fibre Reinforced Concrete, *Cement and Concrete Research*, Vol. 30, No.1, pp. 63-69, ISSN 0008-8846.

Ramakrishnan, V.; Meyer, C.; Naaman, A. ; Zhao, G. ; Fang, L. (1996).Cyclic Behavior, Fatigue Strength, Endurance Limit and Models for Fatigue Behavior of FRC, *in: A. Naaman, H.W. Reinhardt (Eds.), High performance fiber reinforced cement composites, London, E & FN Spon*, ISBN 0419211802.

Singh, S.P.& Kaushik, S.K. (2003). Fatigue Strength of Steel Fibre Reinforced Concrete in Flexure, *Cement and Concrete Composites*, Vol. 25, pp. 779-786, ISSN 0958-9465.

Sun, W.; Mu, R.; Luo, X.; Miao, C. (2002). Effect of Chloride Salt, Freeze–Thaw Cycling and Externally Applied Load on the Performance of the Concrete, Cement and Concrete Research Vol.32, pp. 1859–1864, ISSN 0008-8846.

Yildirim, S. T. (2002). The Investigation of Performance Characteristics of Fiber Reinforced Concrete, PhD Thesis, Firat University, Science Institute, Elazığ, Turkey, p.193, (in Turkish).

Permissions

The contributors of this book come from diverse backgrounds, making this book a truly international effort. This book will bring forth new frontiers with its revolutionizing research information and detailed analysis of the nascent developments around the world.

We would like to thank Dr. Fatih Doğan, for lending his expertise to make the book truly unique. He has played a crucial role in the development of this book. Without his invaluable contribution this book wouldn't have been possible. He has made vital efforts to compile up to date information on the varied aspects of this subject to make this book a valuable addition to the collection of many professionals and students.

This book was conceptualized with the vision of imparting up-to-date information and advanced data in this field. To ensure the same, a matchless editorial board was set up. Every individual on the board went through rigorous rounds of assessment to prove their worth. After which they invested a large part of their time researching and compiling the most relevant data for our readers. Conferences and sessions were held from time to time between the editorial board and the contributing authors to present the data in the most comprehensible form. The editorial team has worked tirelessly to provide valuable and valid information to help people across the globe.

Every chapter published in this book has been scrutinized by our experts. Their significance has been extensively debated. The topics covered herein carry significant findings which will fuel the growth of the discipline. They may even be implemented as practical applications or may be referred to as a beginning point for another development. Chapters in this book were first published by InTech; hereby published with permission under the Creative Commons Attribution License or equivalent.

The editorial board has been involved in producing this book since its inception. They have spent rigorous hours researching and exploring the diverse topics which have resulted in the successful publishing of this book. They have passed on their knowledge of decades through this book. To expedite this challenging task, the publisher supported the team at every step. A small team of assistant editors was also appointed to further simplify the editing procedure and attain best results for the readers.

Our editorial team has been hand-picked from every corner of the world. Their multi-ethnicity adds dynamic inputs to the discussions which result in innovative outcomes. These outcomes are then further discussed with the researchers and contributors who give their valuable feedback and opinion regarding the same. The feedback is then collaborated with the researches and they are edited in a comprehensive manner to aid the understanding of the subject.

Apart from the editorial board, the designing team has also invested a significant amount of their time in understanding the subject and creating the most relevant covers. They scrutinized every image to scout for the most suitable representation of the subject and create an appropriate cover for the book.

The publishing team has been involved in this book since its early stages. They were actively engaged in every process, be it collecting the data, connecting with the contributors or procuring relevant information. The team has been an ardent support to the editorial, designing and production team. Their endless efforts to recruit the best for this project, has resulted in the accomplishment of this book. They are a veteran in the field of academics and their pool of knowledge is as vast as their experience in printing. Their expertise and guidance has proved useful at every step. Their uncompromising quality standards have made this book an exceptional effort. Their encouragement from time to time has been an inspiration for everyone.

The publisher and the editorial board hope that this book will prove to be a valuable piece of knowledge for researchers, students, practitioners and scholars across the globe.

List of Contributors

Somaye Allahvaisi
Department of Entomology and Toxicology, Iran
Faculty of Agriculture, Islamic Azad University of Tehran, Iran
Branch of Sciences & Researches, Iran

Soraia Vilela Borges
Universidade Federal de Lavras, Brazil
Departamento de Ciência dos Alimentos, Brazil

Lyudmila Shibryaeva
N.M. Emanuel Institute of Biochemical Physics, Russia
Russian Academy of Sciences, Russia

Hugo Malon, Jesus Martin and Luis Castejon
University of Zaragoza, Spain

Zulkifli Mohamad Ariff and Azlan Ariffin
Universiti Sains Malaysia,

Suzi Salwah Jikan
Universiti Tun Hussein Onn Malaysia,

Nor Azura Abdul Rahim
Universiti Malaysia Perlis, Malaysia

M. Szanser
Polish Academy of Sciences, Poland
Centre for Ecological Research, Łomianki, Poland

Leonora M. Mattos and Celso L. Moretti
Embrapa Vegetables, Brazil

Marcos D. Ferreira
Embrapa Instrumentation, Brazil

Igor Novák, Anton Popelka and Ivan Chodák
Polymer Institute, Slovak Academy of Science,

Ján Sedliačik
Technical University in Zvolen, Slovakia

Wanda Wadas
Siedlce University of Natural Sciences and Humanities, Poland

Asya Viraneva, Temenuzhka Yovcheva and Georgi Mekishev
University of Plovdiv, Department of Experimental Physics, Bulgaria

Ewa Markiewicz
1Institute of Molecular Physics, Polish Academy of Sciences, Poznań,

Dominik Paukszta and Sławomir Borysiak
Institute of Technology and Chemical Engineering, Poznan University of Technology, Poznań, Poland

Salih Taner Yildirim
Kocaeli University, Civil Engineering Department, Kocaeli, Turkey

Cevdet Emin Ekinci
Fırat University, Construction Education Department, Elazığ, Turkey

Printed in the USA
CPSIA information can be obtained
at www.ICGtesting.com
JSHW011415221024
72173JS00004B/552